**Man's Impact on the
Global Environment**

**Assessment and
Recommendations
for Action**

The MIT Press
Cambridge, Massachusetts,
and London, England

Man's Impact on the Global Environment

Assessment and
Recommendations
for Action

Report of the Study of Critical
Environmental Problems
(SCEP)

Sponsored by the Massachusetts
Institute of Technology

ISBN 0 262 19086 9 (hardcover)

ISBN 0 262 69027 6 (paperback)

Library of Congress catalog card number: 74–139447

Contents

Preface

The need for the 1970 Study of Critical Environmental Problems (SCEP) was perceived in June 1969 in discussions we had with a number of scientists and public officials. In examining the status of governmental and nongovernmental preparations for the 1972 United Nations Conference on the Human Environment, several of us concluded that an initiative such as this Study would provide an important input into planning for that Conference and for numerous other national and international activities.

The Steering Committee that planned the Study during the fall of 1969 and spring of 1970 was chaired by Carroll L. Wilson, Professor of Management, M.I.T., and included Raymond F. Baddour, Chairman, Department of Chemical Engineering, M.I.T.; Raymond L. Bisplinghoff, Dean, School of Engineering, M.I.T.; John L. Buckley, Office of Science and Technology, Executive Office of the President; Richard A. Carpenter, Chief, Environmental Policy Division, Legislative Reference Service of the Library of Congress; Paul M. Fye, Director, Woods Hole Oceanographic Institution; Thomas F. Malone, Professor of Physics, University of Connecticut; William H. Matthews, Department of Political Science, M.I.T.; Richard S. Morse, Senior Lecturer, School of Management, M.I.T.; George W. Rathjens, Professor of Political Science, M.I.T.; and Roger Revelle, Director, Center for Population Studies, Harvard University.

By mid-November 1969, the Steering Committee had chosen as the topics for the Study those problems arising from the impact of man's activities on the global environment. Unlike many environmental problems of more immediate or local concern, global problems such as changes in climate and in ocean and terrestrial ecosystems had not been subjected to intensive study and examination. SCEP was developed to fill that gap.

The Steering Committee felt that a one-month, interdisciplinary study of these complex problems could provide citizens, public policy makers, and scientists with an authoritative assessment of the degree and nature of man's impact on the global environment and with specific recommendations for new programs of focused research, monitoring, and action. It was also hoped that if such a multidisciplinary and systematic study of a specific set of global problems proved effective in raising the level of informed public and scientific discussion of, and action on, these issues this Study might serve as a model for similar multidisciplinary efforts that could attack many of the other critical problems of our time.

During the winter and spring of 1969–1970, support for the Study was sought and obtained, participants were invited, and extensive background preparations were undertaken. Materials assembled for SCEP included approximately two hundred papers and articles of which about one-fourth were prepared especially for the Study.

Approximately forty scientists and professionals participated in SCEP for almost the entire month of July 1970. In addition, about thirty other part-time participants attended for periods ranging from one to three weeks and made substantive contributions to one or more SCEP Work Groups. The Study was also greatly aided by the work of another forty-five scientists and professionals who attended for varying lengths of time as consultants or observers.

SCEP participants represented expertise in over a dozen disciplines, including meteorology, atmospheric chemistry, oceanography, biology, ecology, geology, physics, several branches of engineering, economics, social sciences, and law; and they were drawn from seventeen universities, thirteen federal departments and agencies, three national laboratories, and eleven nonprofit and industrial corporations. The research and rapporteurial staff of the Study included eleven graduate and law students from three universities.

The Study, which was sponsored by the Massachusetts Institute of Technology (M.I.T.), concentrated on the global climatic and ecological effects of several specific pollutants in the atmosphere-land-ocean system. This Report presents the scientific and technical judgments of the participants on the present status of understanding of several global problems. In addition, SCEP explored the procedures and programs of focused research, monitoring, and action that will be required to understand further the nature of potential threats to the global environment so that effective action can be taken to avert future crises. The specific recommendations of SCEP are outlined in this Report.

The Study began on Wednesday, July 1, 1970, at Williams College in Williamstown, Massachusetts. During the first two days, several participants made presentations to the assembled group that established the framework for the month-long effort. By Friday, Work Groups for the first full week of study had been identified. These five Work Groups, which met during the first week (July 7–12), were concerned with evaluating present knowledge of the rates, routes, and reservoirs of several pollutants that might have harmful global effects. Two of the groups were concerned with routes and reservoirs in the atmosphere and the oceans. The other three groups developed the data base for the Study by determining the sources and rates of relevant pollutants from three major sectors of man's activities—industrial wastes; domestic, agricultural, and mining wastes; and energy products.

During the second full week of the Study (July 13–19), the participants divided into four Work Groups: Climatic Effects, Ecological Effects (on ocean and terrestrial ecosystems), Monitoring, and Implications of Change. These Work Groups continued work into the third week, and then the participants mixed to some degree to develop recommendations for programs of focused research, monitoring, and action. The first three days of the fourth week (July 27–29) were devoted to the adoption of the individual Work Group reports, and during the final two days extensive briefings were given to the supporters of SCEP and to the news media.

This Report is divided into two major parts. The first part is a distillation of the major findings and recommendations developed by the various Work Groups of the Study. This part is, in a broad sense, the SCEP Report. All those attending the Study had an opportunity to examine these conclusions and the papers on which they were based, but all participants had neither the time nor the expertise required to make an independent judgment on each and every area discussed in the first part. Therefore, it should not be assumed that each Study participant subscribes to every statement in this Report.

The second part of this Report contains the reports of seven SCEP Work Groups (several first- and second-week Work Groups combined to write single reports). These Work Group reports were developed through intensive, full-time discussion and study by the group members. In some cases these deliberations continued for the entire Study period. These reports represent the consensus of those whose names appear on the individual Work Group reports. Each member concurs with the substantive presentation, but not all members had an opportunity to review the final wording of the reports. Participants in the Study acted as individuals, not as representatives of the agencies or organizations with which they were affiliated.

Because the major objective of SCEP was to raise the level of informed public and scientific discussion and action on global environmental problems, this Report was published as quickly as possible. Rapid publication has undoubtedly resulted in less editorial smoothness than might have been possible if individual Work Group reports had been substantially reworked following the Study. These reports reflect the collective judgments of group members, and it was felt that extensive editorial work might alter the meanings and emphases of important points. Therefore, the Work Group reports appear in the book in essentially the same forms that were reviewed and adopted by the groups' members at the end of July.

The reports of Work Groups 5, 6, and 7, which comprise the data base of the Study, are exceptions to this statement. These groups were in existence only for the first week of the Study, and their members compiled many of the data and references that are cited in the reports of this book. However, during the remainder of the Study and for a short period thereafter, these data were carefully examined by Raymond F. Baddour and several participants, consultants, and Study staff. Work Group reports 5, 6, and 7 are generally shorter and less extensive than those initially compiled, for they include only data pertinent to the other Work Group reports and data for which documentation could be readily verified.

A series of edited volumes will be published by The M.I.T. Press which will include many of the working and background papers produced by or for SCEP. Comprised of papers signed by the individuals or small groups which produced them, those volumes will provide the substantive support for the Work Group reports contained in this book.

The following federal departments and agencies, private

foundations, and organizations supported the Study and follow-up activities through preparation of background materials and professional participation in the Study and through grants or contracts:

Agricultural Research Service; Department of Agriculture

Atomic Energy Commission

Department of State

Department of Transportation

Environmental Science Services Administration; Department of Commerce

Forest Service; Department of Agriculture

National Aeronautics and Space Administration

National Air Pollution Control Administration; Department of Health, Education, and Welfare

National Science Foundation

American Conservation Association

Ford Foundation

Rockefeller Foundation

Sloan Foundation

Center for the Environment and Man, Inc.

Massachusetts Institute of Technology

In addition, the following federal departments and agencies and organizations provided support for SCEP through preparation of background materials and professional participation:

Coast Guard; Department of Transportation

Federal Water Quality Administration; Department of Interior

Geological Survey; Department of Interior

National Academy of Sciences

National Center for Atmospheric Research

Oak Ridge National Laboratory

RAND Corporation

Allied Chemical Corporation

American Electrical Power Company, Inc.

Chemical Construction Corporation; Boise Cascade Corporation

Consolidated Edison Corporation of New York

Esso Research & Engineering Company

General Electric Company

Without the extraordinary support provided by these organizations and many persons within them and by numerous other individuals, the Study could not have developed a Report of this scope and depth. On behalf of the Study participants and the Steering Committee, we wish to thank all those who contributed to this effort.

Carroll L. Wilson, SCEP Director
William H. Matthews, Associate Director

Participants

Director
Carroll L. Wilson
SLOAN SCHOOL OF MANAGEMENT
AND
PROGRAM FOR THE SOCIAL APPLI-
CATION OF TECHNOLOGY
MASSACHUSETTS INSTITUTE OF
TECHNOLOGY

Associate Director
William H. Matthews
DEPARTMENT OF CIVIL ENGINEER-
ING, AND
PROGRAM FOR THE SOCIAL APPLI-
CATION OF TECHNOLOGY
MASSACHUSETTS INSTITUTE OF
TECHNOLOGY

Geirmundur Arnason
CENTER FOR THE ENVIRONMENT
AND MAN, INC.

Robert U. Ayres
INTERNATIONAL RESEARCH AND
TECHNOLOGY CORPORATION

Raymond F. Baddour, Head
DEPARTMENT OF CHEMICAL
ENGINEERING
MASSACHUSETTS INSTITUTE OF
TECHNOLOGY

John F. Brown, Jr., Manager
LIFE SCIENCES BRANCH
GENERAL ELECTRIC RESEARCH AND
DEVELOPMENT CENTER

Richard D. Cadle, Head
CHEMISTRY AND MICROPHYSICS
DEPARTMENT
NATIONAL CENTER FOR ATMOS-
PHERIC RESEARCH

Robert Citron, Director
CENTER FOR SHORT-LIVED
PHENOMENA
THE SMITHSONIAN INSTITUTION

Seymour Edelberg
LINCOLN LABORATORY
MASSACHUSETTS INSTITUTE OF
TECHNOLOGY

Gifford Ewing
WOODS HOLE OCEANOGRAPHIC
INSTITUTION

John Franklin
GREENWICH, CONNECTICUT

Edward D. Goldberg
SCRIPPS INSTITUTION OF
OCEANOGRAPHY

M. Grant Gross
MARINE SCIENCES RESEARCH
CENTER
STATE UNIVERSITY OF NEW YORK
AT STONY BROOK

Edward Hamilton
THE BROOKINGS INSTITUTION

Bruce B. Hanshaw, Deputy
Assistant Director
U.S. GEOLOGICAL SURVEY

J. B. Hilmon, Chief
BRANCH OF RANGE AND WILDLIFE
HABITAT ECOLOGY AND MANAGE-
MENT RESEARCH
U.S. FOREST SERVICE

Dale W. Jenkins, Director
OFFICE OF ECOLOGY
THE SMITHSONIAN INSTITUTION

Milton Katz
HARVARD LAW SCHOOL

Philip C. Kearny, Investiga-
tions Leader
PESTICIDE BEHAVIOR IN SOILS
AGRICULTURAL RESEARCH SERVICE

Charles D. Keeling
SCRIPPS INSTITUTION OF
OCEANOGRAPHY

William W. Kellogg, Associate
Director
NATIONAL CENTER FOR ATMOS-
PHERIC RESEARCH

Jules Lehman
NATIONAL AERONAUTICS AND
SPACE ADMINISTRATION

Julius London
DEPARTMENT OF ASTRO-
GEOPHYSICS
UNIVERSITY OF COLORADO

Frank G. Lowman, Head
RADIOECOLOGY DIVISION
PUERTO RICO NUCLEAR CENTER

Lester Machta, Director
AIR RESOURCES LABORATORY
ENVIRONMENTAL SCIENCE SERV-
ICES ADMINISTRATION

William H. Matthews
DEPARTMENT OF CIVIL ENGINEER-
ING
MASSACHUSETTS INSTITUTE OF
TECHNOLOGY

Reginald E. Newell
DEPARTMENT OF METEOROLOGY
MASSACHUSETTS INSTITUTE OF
TECHNOLOGY

Jerry S. Olson
HEALTH PHYSICS DIVISION
OAK RIDGE NATIONAL
LABORATORY

Hans A. Panofsky
PROFESSOR OF ATMOSPHERIC
SCIENCES
PENNSYLVANIA STATE UNIVERSITY

James T. Peterson
NATIONAL AIR POLLUTION CON-
TROL ADMINISTRATION

Walter Ramberg
INTERNATIONAL SCIENTIFIC AND
TECHNOLOGICAL AFFAIRS
DEPARTMENT OF STATE

George W. Rathjens
DEPARTMENT OF POLITICAL
SCIENCE
MASSACHUSETTS INSTITUTE OF
TECHNOLOGY

Henry Reichle
LANGLEY RESEARCH CENTER
NATIONAL AERONAUTICS AND
SPACE ADMINISTRATION

Joseph Reid
SCRIPPS INSTITUTION OF
OCEANOGRAPHY

G. D. Robinson
CENTER FOR THE ENVIRONMENT
AND MAN, INC.

Silvio Simplicio, Director
WEATHER BUREAU
EASTERN REGION

Frederick E. Smith
GRADUATE SCHOOL OF DESIGN
HARVARD UNIVERSITY

Walter O. Spofford
QUALITY OF THE ENVIRONMENT
PROGRAM
RESOURCES FOR THE FUTURE

Howard J. Taubenfeld
SCHOOL OF LAW
SOUTHERN METHODIST
UNIVERSITY

Rita F. Taubenfeld
INSTITUTE OF AERO-SPACE LAW
SOUTHERN METHODIST
UNIVERSITY

Morris Tepper
DEPUTY DIRECTOR OF EARTH
OBSERVATION PROGRAM AND
DIRECTOR OF METEOROLOGY
NATIONAL AERONAUTICS AND
SPACE ADMINISTRATION

F. Joachim Weyl
DEAN OF SCIENCES AND MATHE-
MATICS
HUNTER COLLEGE OF THE CITY
UNIVERSITY OF NEW YORK

Carroll L. Wilson
SLOAN SCHOOL OF MANAGEMENT
MASSACHUSETTS INSTITUTE OF
TECHNOLOGY

George M. Woodwell
BIOLOGY DEPARTMENT
BROOKHAVEN NATIONAL
LABORATORY

Part-Time Participants

A. P. Altshuller, Director
DIVISION OF CHEMISTRY AND
PHYSICS
NATIONAL AIR POLLUTION CON-
TROL ADMINISTRATION

Neil R. Andersen, Chief
CHEMICAL OCEANOGRAPHY
BRANCH
U.S. COAST GUARD

Norman H. Brooks
DEPARTMENT OF CIVIL ENGINEER-
ING
CALIFORNIA INSTITUTE OF TECH-
NOLOGY

Philip A. Butler, Research
Consultant
BUREAU OF COMMERCIAL FISH-
ERIES
U.S. FISH AND WILDLIFE SERVICE

E. W. Callahan, General
Manager
ENVIRONMENTAL SERVICES
DEPARTMENT
ALLIED CHEMICAL CORPORATION

Richard A. Carpenter, Chief
ENVIRONMENTAL POLICY DIVISION
LEGISLATIVE REFERENCE SERVICE
LIBRARY OF CONGRESS

Edward Corino
ESSO RESEARCH & ENGINEERING
COMPANY

Willard A. Crandall, Chief
Chemical Engineer
CONSOLIDATED EDISON COMPANY
OF NEW YORK, INC.

Robert G. Fleagle, Chairman
DEPARTMENT OF ATMOSPHERIC
SCIENCES
UNIVERSITY OF WASHINGTON

Stanley Greenfield, Head
DEPARTMENT OF ENVIRONMENTAL
SCIENCES
RAND CORPORATION

Arthur D. Hasler
LABORATORY OF LIMNOLOGY
UNIVERSITY OF WISCONSIN

Christian Junge, Director
MAX-PLANCK INSTITUT FÜR
CHEMIE
MAINZ, GERMANY

Henry J. Kellermann
Executive Secretary
COMMITTEE ON INTERNATIONAL
ENVIRONMENTAL PROGRAMS
NATIONAL ACADEMY OF SCIENCES

Bostwick Ketchum, Associate
Director
WOODS HOLE OCEANOGRAPHIC
INSTITUTION

Allen V. Kneese, Director
QUALITY OF ENVIRONMENT
PROGRAM
RESOURCES FOR THE FUTURE, INC.

John H. Ludwig
ASSISTANT COMMISSIONER FOR
SCIENCE AND TECHNOLOGY
NATIONAL AIR POLLUTION CON-
TROL ADMINISTRATION

Thomas Marqueen
CHEMICAL CONSTRUCTION COR-
PORATION
BOISE CASCADE CORPORATION

Paul Meier
DEPARTMENT OF STATISTICS
UNIVERSITY OF CHICAGO

David W. Menzel
WOODS HOLE OCEANOGRAPHIC
INSTITUTION

J. Murray Mitchell, Jr.
ENVIRONMENTAL SCIENCE SERV-
ICES ADMINISTRATION

George B. Morgan, Director
DIVISION OF AIR QUALITY AND
EMISSION DATA
NATIONAL AIR POLLUTON CON-
TROL ADMINISTRATION

Robert Risebrough
INSTITUTE OF MARINE RESOURCES
UNIVERSITY OF CALIFORNIA
BERKELEY

Hilliard Roderick
DIRECTORATE FOR SCIENTIFIC
AFFAIRS
ORGANIZATION FOR ECONOMIC
COOPERATION AND DEVELOPMENT

Joseph Smagorinsky
GEOPHYSICAL FLUID DYNAMICS
LABORATORY
ENVIRONMENTAL SCIENCE SERV-
ICES ADMINISTRATION
PRINCETON UNIVERSITY

Lucille F. Stickel
PATUXENT WILDLIFE RESEARCH
CENTER
DEPARTMENT OF THE INTERIOR

Clarence Tarzwell, Director
NATIONAL MARINE WATER QUAL-
ITY LABORATORY
FEDERAL WATER QUALITY ADMIN-
ISTRATION

Herbert L. Volchok
ATOMIC ENERGY COMMISSION

Edward Wenk, Jr.
PROFESSOR OF ENGINEERING AND
PUBLIC AFFAIRS
UNIVERSITY OF WASHINGTON

Howard Wiedemann
DEPARTMENT OF STATE

Consultants and Observers

Arthur Alexiou
SEA GRANT PROGRAM
NATIONAL SCIENCE FOUNDATION

R. Keith Arnold, Deputy Chief
of Forestry Research
U.S. FOREST SERVICE

Charles C. Bates
SCIENCE ADVISOR TO THE COM-
MANDANT
U.S. COAST GUARD

David Beckler,
Assistant to the Director
OFFICE OF SCIENCE AND TECH-
NOLOGY
EXECUTIVE OFFICE OF THE PRESI-
DENT

Nyle C. Brady, Director
CORNELL UNIVERSITY EXPERI-
MENT STATION

Douglas Brooks, Executive
Secretary
EXECUTIVE COUNCIL
OFFICE OF THE DIRECTOR
NATIONAL SCIENCE FOUNDATION

Reid Bryson
DEPARTMENT OF METEOROLOGY
UNIVERSITY OF WISCONSIN

John L. Buckley
OFFICE OF SCIENCE AND TECH-
NOLOGY
EXECUTIVE OFFICE OF THE PRESI-
DENT

Theodore C. Byerly
ASSISTANT DIRECTOR OF SCIENCE
AND EDUCATION
OFFICE OF THE SECRETARY
DEPARTMENT OF AGRICULTURE

Alan Carlin
ECONOMICS DEPARTMENT
RAND CORPORATION

William Drury
MASSACHUSETTS AUDUBON
SOCIETY

Leonard Dworsky, Director
WATER RESOURCES AND MARINE
SCIENCES CENTER
CORNELL UNIVERSITY

Robert Ellis, President
CENTER FOR THE ENVIRONMENT
AND MAN, INC.

Gordon Everett
ACTING DEPUTY ASSISTANT
SECRETARY
WATER QUALITY AND RESEARCH
DEPARTMENT OF THE INTERIOR

Hermann Feltz
U.S. GEOLOGICAL SURVEY

T. T. Frankenberg
CONSULTING MECHANICAL
ENGINEER
AMERICAN ELECTRIC POWER COM-
PANY, INC.

Paul M. Fye, Director
WOODS HOLE OCEANOGRAPHIC
INSTITUTION

Milton Harris, Chairman of
the Board
AMERICAN CHEMICAL SOCIETY

Wilmot Hess
RESEARCH LABORATORIES
ENVIRONMENTAL SCIENCE SERV-
ICES ADMINISTRATION

Alan Hoffman
DIVISION OF AIR QUALITY AND
EMISSION DATA
NATIONAL AIR POLLUTION CON-
TROL ADMINISTRATION

Everett L. Hollis
MAYER, BROWN, AND PLATT

Merton Ingham, Chief
OCEANOGRAPHY UNIT
U.S. COAST GUARD

Claus Ludwig
CONVAIR DIVISION
GENERAL DYNAMICS CORPORA-
TION

Thomas Malone
DEPARTMENT OF PHYSICS
SPECIAL CONSULTANT TO THE
PRESIDENT ON ENVIRONMENTAL
PROBLEMS
UNIVERSITY OF CONNECTICUT

Robert A. McCormick, Director
DIVISION OF METEOROLOGY
NATIONAL AIR POLLUTION CON-
TROL ADMINISTRATION

Richard S. Morse
SLOAN SCHOOL OF MANAGEMENT
MASSACHUSETTS INSTITUTE OF
TECHNOLOGY

Vaun A. Newill, Director
DIVISION OF HEALTH EFFECTS
RESEARCH
NATIONAL AIR POLLUTION CON-
TROL ADMINISTRATION

Conrad Neuman, Physical Sci-
ence Administrator
ENVIRONMENTAL SCIENCE DIVI-
SION
NATIONAL SCIENCE FOUNDATION

Gaylord Northrup
CENTER FOR THE ENVIRONMENT
AND MAN, INC.

Captain R. I. Price, Chief
PLANNING AND SPECIAL PROJECTS
STAFF
OFFICE OF MERCHANT MARINE
SAFETY
U.S. COAST GUARD

Martin Prochnik
DEPUTY SCIENCE ADVISOR TO THE
SECRETARY
DEPARTMENT OF THE INTERIOR

Robert H. Quig, Manager
POLLUTION CONTROL DIVISION
CHEMICAL CONSTRUCTION COR-
PORATION
BOISE CASCADE CORPORATION

Roger Revelle, Director
CENTER FOR POPULATION STUDIES
HARVARD UNIVERSITY

Harry Richardson
CHEMICAL CONSTRUCTION COR-
PORATION
BOISE CASCADE CORPORATION

William Salmon
INTERNATIONAL SCIENTIFIC AND
TECHNOLOGICAL AFFAIRS
DEPARTMENT OF STATE

Vincent Schaefer
STATE UNIVERSITY OF NEW YORK
AT ALBANY

Herbert A. Simon
PSYCHOLOGY AND COMPUTER
SCIENCE
CARNEGIE-MELLON UNIVERSITY

Norman Sohl
U.S. GEOLOGICAL SURVEY

Kurt Stehling
NATIONAL MARINE RESOURCES
COUNCIL
EXECUTIVE OFFICE OF THE PRESI-
DENT

Lyle Tiffany
BENDIX CORPORATION

Thomas Winter, Staff Member
COUNCIL ON ENVIRONMENTAL
QUALITY
EXECUTIVE OFFICE OF THE PRESI-
DENT

Harry G. Woodbury
CONSOLIDATED EDISON COMPANY
OF NEW YORK, INC.

Harold Yates, Director
SATELLITE EXPERIMENTAL LAB-
ORATORY
ENVIRONMENTAL SCIENCE SERV-
ICES ADMINISTRATION

Research Staff
Ruben S. Brown, SCEP Staff
Coordinator
MASSACHUSETTS INSTITUTE OF
TECHNOLOGY

Gregory Arenson
MASSACHUSETTS INSTITUTE OF
TECHNOLOGY

Steven Carhart
MASSACHUSETTS INSTITUTE OF
TECHNOLOGY

Ada Demb
HARVARD UNIVERSITY

Herbert Quinn
HARVARD UNIVERSITY

Rapporteurs
William G. Anderson
HARVARD LAW SCHOOL

Stephen B. Burbank
HARVARD LAW SCHOOL

Jonathan Marks
HARVARD LAW SCHOOL

Peter Katz
HARVARD LAW SCHOOL

Terry Schaich
FLETCHER SCHOOL OF LAW AND
DIPLOMACY

Robert E. Stoller
HARVARD LAW SCHOOL

**Arrangements
Director**
Mary Kathleen Brown
HARVARD UNIVERSITY

Part I
Summary of Findings and Recommendations

1.
Introduction

1.1

The Problems Studied

Over the past few years, the concept of the earth as a "spaceship" has provided many people with an awareness of the finite resources and the complex natural relationships on which man depends for his survival. These realizations have been accompanied by concerns about the impacts that man's activities are having on the global environment. Some concerned individuals, including well-known scientists, have warned of both imminent and potential global environmental catastrophes.

Theories and speculations of the global effects of pollution have included assertions that the buildup of CO_2 from fossil fuel combustion might warm up the planet and cause the polar ice to melt, thus raising the sea level several hundred feet and submerging coastal cities. Equally foreboding has been the warning of the possibility that particles emitted into the air from industrial, energy, and transportation processes might prevent some sunlight from reaching the earth's surface, thus lowering global temperature and beginning a new ice age. Demands to ban DDT have been increasing steadily as its effects on the reproductive capabilities of birds have been determined, and as evidence is found of its accumulation in other species including man. Serious questions have been raised about the effects on ocean and terrestrial ecosystems of systematically discharging into the environment such toxic materials as heavy metals, oil, and radioactive substances; or of nutrients such as phosphorus which can overenrich lakes and coastal areas.

This Report of the one-month, interdisciplinary Study of Critical Environmental Problems (SCEP) presents an assessment of the existing state of scientific knowledge on these and related global environmental problems and contains specific recommendations for action which would reduce the harmful effects of pollution or would provide the information required to understand more adequately the impact of man on the global environment. If such information is not obtained, some critical environmental questions will remain unresolved and we may never be able to identify potential crises in enough time to avoid them and possibly to prevent irreversible global damage.

For each of these global problem areas, Work Groups of the Study addressed the following questions:

What can we now authoritatively say on the subject?

What are the gaps in knowledge which limit our confidence in the assessments we can now make?

What must be done to improve the data and our understanding of their significance so that better assessments may be made in the future?

What programs of focused research, monitoring, and action are needed?

What are the characteristics of the national and/or international action needed to implement the recommendations of the Study?

1.2
The Focus on Global Problems

In order to use most effectively the resources and time available for this Study, it was necessary to limit the scope and character of the problems that were chosen for intensive investigation. SCEP focused on environmental problems whose cumulative effects on ecological systems are so large and prevalent that they have worldwide significance. Thus the Study was primarily concerned with the effects of pollution on man through changes in climate, ocean ecology, or in large terrestrial ecosystems.

In general, local and regional environmental problems, the first-order effects of population growth, and the direct health effects of pollution on man were not considered by the Study. This choice does not imply that these latter problem areas are not of critical concern. Indeed, they are so important that many organizations are deeply concerned with studying and ameliorating them. However, no organization is charged with the responsibility for determining the status of the total global environment and alerting man to dangers that may result from his practices. SCEP attempted to perform this function.

It should be noted that the existence of a global problem does not imply the necessity for a global solution. The sources of pollution are activities of man that can often be effectively controlled or regulated where they occur. Most corrective action will probably have to be taken at the national, regional, and local levels. In research and monitoring programs, however, the potential for international cooperation is high. Effective cooperation now might increase the likelihood of smooth international relations should a global problem ever demand strict international regulation or control of pollution-producing activities. In discussing global environmental problems, it is also necessary to consider the different perspectives of highly industrialized and developing countries regarding pollution.

1.3
The Quality of Available Data and Projections

Before discussing the findings, conclusions, and recommendations of the Study, it is important to note the deficiencies in the data and projections related to problems of global concern. In the process of making judgments we found that critically needed data were fragmentary, contradictory, and in some cases completely unavailable. This was true for all types of data—scientific, technical, economic, industrial, and social. These conditions existed despite a year of planning, extensive preparation of background materials, the presence among Study participants of some of the world's leading scientists, and the generous access to data provided by virtually every relevant federal agency.

With respect to economic and industrial statistical data and projections needed to determine trends of environmental contamination, we found firm data only up to 1967 or 1968 for the United States. International compilations of such data are often incomplete and are of questionable reliability because of uncertainties and inconsistencies in reporting, and because of lack of mechanisms to verify or standardize reports of cooperating nations. Very few projections exist for rates of growth of various industrial sectors, relevant domestic and agricultural activities of man, and energy demands. Those that are available are often based on different and sometimes questionable assumptions and methodologies.

Data on important physical, chemical, and ecological phenomena and parameters are also inadequate for providing the foundation for definitive statements about environmental effects. Specific recommendations for obtaining these data appear throughout this Report. The present data base for global problems is so poor, however, that three general recommendations are necessary.

Recommendations

1.   We recommend the development of new methods for gathering and compiling global economic and statistical information, which organize data across traditional areas of environmental responsibility, such as air and water pollution. We further recommend the propagation of uniform data-collection standards to ensure, for example, that industrial production data collection across the world will be of comparable precision and focus.

2.   We recommend a study of the possibility of setting up international physical, chemical, and ecological measurement standards, to be administered through a monitoring standards center with a "real time" data analysis capability, allowing for prompt feedback to monitoring units in terms of monitoring or measurement parameters, levels of accuracy, frequency of observation, and other factors.

3.   We recommend an immediate study of global monitoring to examine the scientific and political feasibility of integration of existing and planned monitoring programs and to set out steps necessary to establish an optimal system.

2.
Climatic Effects of Man's Activities

2.1

Introduction

There is geological evidence that there have been five or six glacial periods (ice ages); the most recent (the Pleistocene) lasted 1 to 1.5 million years. In the past century there has been a general warming of the atmosphere of about 4°C up to 1940, followed by a few tenths degree cooling. It seems clear that our climate is subject to a wide variety of fluctuations, with periods ranging from decades to millennia, and that it is changing now.

We know that the atmosphere is a relatively stable system. The solar radiation that is absorbed by the planet and heats it must be almost exactly balanced by the emitted terrestrial infrared radiation that cools it; otherwise the mean temperature would change much more rapidly than just noted. This nearly perfect balance is the key to the changes that do occur, since a reduction of only about 2 percent in the available energy can, in theory, lower the mean temperature by 2°C and produce an ice age.

That there have not been wider fluctuations in climate is our best evidence that the complex system of ocean and air currents, evaporation and precipitation, surface and cloud reflection and absorption form a complex feedback system for keeping the global energy balance nearly constant. Nonetheless, the delicacy of this balance and the consequences of disturbing it make it very important that we attempt to assess the present and prospective impact of man's activities on this system.

The total mass of the atmosphere and the energy involved in even such a minor disturbance as a thunderstorm (releasing the energy equivalent to many hydrogen bombs) should convince us immediately that man cannot possibly hope to intervene in such a gigantic arena. However, in reality man does intervene, because he can—without intending to do so—reach some leverage points in the system.

All the important leverage points that this Study has identified control the radiation balance of the atmosphere in one way or another, and most of them control it by changing the composition of the atmosphere. For example, man can change the temperature of the atmosphere by introducing a gas such as CO_2 or a cloud of particles that absorbs and emits solar and terrestrial infrared radiation, thereby altering the delicate balance we have described. He can also affect the heat balance by changing the face of the earth or by adding heat as a result of rising energy demands.

A thorough understanding and reliable prediction of the influence of atmospheric pollutants on climate requires the mathematical simulation of atmosphere-ocean systems, including the pollutants. At present, computer models successfully simulate many observed characteristics of the climate and have significantly advanced our knowledge of atmospheric phenomena. They have, however, a number of drawbacks that become serious when modeling new states of equilibrium or changes of climate in its transition toward these new states. Unless these limitations are overcome, it will be difficult, if not impossible, to predict inadvertent climate modifications that might be caused by man.

Recommendations
1.   We recommend that current computer models be improved by including more realistic simulations of clouds and air-sea interaction and that attempts be made to include particles when their properties become better known. Such models should be run for periods of at least several simulated years. The effects of potential global pollutants on the climate and on phenomena such as cloud formation should be studied with these models.
2.   We recommend that possibilities be investigated for simplifying existing models to provide a better understanding of climatic changes. Simultaneously, a search should be made for alternative types of models which are more suitable for handling problems of climatic change.

2.2
Carbon Dioxide from Fossil Fuels
All combustion of fossil fuels produces carbon dioxide (CO_2), which has been steadily increasing in the atmosphere at 0.2 percent per year since 1958. Half of the amount man puts into the atmosphere stays and produces this rise in concentration. The other half goes into the biosphere and the oceans, but we are not certain how it is divided between these two reservoirs. CO_2 from fossil fuels is a small part of the natural CO_2 that is constantly being exchanged between the atmosphere/oceans and the atmosphere/forests.

A projected 18 percent increase resulting from fossil fuel

combustion to the year 2000 (from 320 ppm to 379 ppm) might increase the surface temperature of the earth 0.5°C; a doubling of the CO_2 might increase mean annual surface temperatures 2°C. This latter change could lead to long-term warming of the planet. These estimates are based on a relatively primitive computer model, with no consideration of important motions in the atmosphere, and hence are very uncertain. However, these are the only estimates available today.

Should man ever be compelled to stop producing CO_2, no coal, oil, or gas could be burned and all industrial societies would be drastically affected. The only possible alternative for energy for industrial and commercial use is nuclear energy, whose by-products may also cause serious environmental effects. There are at present no electric motor vehicles that could be used on the wide scale our society demands.

Although we conclude that the probability of direct climate change in this century resulting from CO_2 is small, we stress that the long-term potential consequences of CO_2 effects on the climate or of societal reaction to such threats are so serious that much more must be learned about future trends of climate change. Only through these measures can societies hope to have time to adjust to changes that may ultimately be necessary.

Recommendations

1.   We recommend the improvement of present estimates of future combustion of fossil fuels and the resulting emissions.

2.   We recommend study of changes in the mass of living matter and decaying products.

3.   We recommend continuous measurement and study of the carbon dioxide content of the atmosphere in a few areas remote from known sources for the purpose of determining trends. Specifically, four stations and some aircraft flights are required.

4.   We recommend systematic scientific study of the partition of carbon dioxide among the atmosphere, the oceans, and the biomass. Such research might require up to 12 stations.

2.3

Particles in the Atmosphere

Fine particles change the heat balance of the earth because they both reflect and absorb radiation from the sun and the earth.

Large amounts of such particles enter the troposphere (the zone up to about 12 km or 40,000 feet) from natural sources such as sea spray, windblown dust, volcanoes, and from the conversion of naturally occurring gases into particles.

Man introduces fewer particles into the atmosphere than enter from natural sources; however, he does introduce significant quantities of sulfates, nitrates, and hydrocarbons. The largest single artificial source is the production of sulfur dioxide from the burning of fossil fuel that subsequently is converted to sulfates by oxidation. Particle levels have been increasing over the years as observed at stations in Europe, North America, and the North Atlantic but not over the Central Pacific.

In the troposphere, the residence times of particles range from 6 days to 2 weeks, but in the lower stratosphere micronsize particles or smaller may remain for 1 to 3 years. This long residence time in the stratosphere and also the photochemical processes occurring there make the stratosphere more sensitive to injection of particles than the troposphere.

Particles in the troposphere can produce changes in the earth's reflectivity, cloud reflectivity, and cloud formation. The magnitudes of these effects are unknown, and in general it is not possible to determine whether such changes would result in a warming or cooling of the earth's surface. The area of greatest uncertainty in connection with the effects of particles on the heat balance of the atmosphere is our current lack of knowledge of their optical properties in scattering or absorbing radiation from the sun or the earth.

Particles also act as nuclei for condensation or freezing of water vapor. Precipitation processes can certainly be affected by changing nuclei concentrations, but we do not believe that the effect of man-made nuclei will be significant on a global scale.

Recommendations
1.   We recommend studies to determine optical properties of fine particles, their sources, transport processes, nature, size distributions, and concentrations in both the troposphere and stratosphere, and their effects on cloud reflectivity.
2.   We recommend that the effects of particles on radiative transfer be studied and that the results be incorporated in mathe-

matical models to determine the influence of particles on planetary circulation patterns.

3.   We recommend extending and improving solar radiation measurements.

4.   We recommend beginning measurements by lidar (optical radar) methods of the vertical distribution of particles in the atmosphere.

5.   We recommend the study of the scientific and economic feasibility of initiating satellite measurements of the albedo (reflectivity) of the whole earth, capable of detecting trends of the order of 1 percent per 10 years.

6.   We recommend beginning a continuing survey, with ground and aircraft sampling, of the atmosphere's content of particles and of those trace gases that form particles by chemical reactions in the atmosphere. For relatively long-lived constituents about 10 fixed stations will be required, for short-lived constituents, about 100.

7.   We recommend monitoring several specific particles and gases by chemical means. About 100 measurement sites will be required.

2.4
The Role of Clouds

The importance of clouds in the atmosphere stems from their relatively high reflectivity for solar radiation and their central role in the various processes involved in the heat budget of the earth-atmosphere system.

Recommendations

1.   We recommend that there be global observations of cloud distribution and temporal variations. High spatial resolution satellite observations are required to give "correct" cloud population counts and to establish the existence of long-term trends in cloudiness (if there are any).

2.   We recommend studies of the optical (visible and infrared) properties of clouds as functions of the various relevant cloud and impinging radiation parameters. These studies should include the effect of particles on the reflectivity of clouds and a determination of the infrared "blackness" of clouds.

2.5
Cirrus Clouds from Jet Aircraft

Contrail (condensation trail) formation, which is common near the world's air routes, is more likely to occur when jets fly in the upper troposphere than in the lower troposphere because of the different meteorological conditions in these two regions.

There are very few, if any, statistics that permit us to determine whether the advent of commercial jet aircraft has altered the frequency of occurrence or the properties of cirrus clouds. We do not know whether the projected increase in the operation of subsonic jets will have any climate effects.

Two weather effects from enhanced cirrus cloudiness are possible. First, the radiation balance may be slightly upset, and, second, cloud seeding by falling ice crystals might initiate precipitation sooner than it would otherwise occur.

Recommendations

1.   We recommend that the magnitude and distribution of increased cirrus cloudiness from subsonic jet operations in the upper troposphere be determined. A study of the phenomenon should be conducted by examining cloud observations at many weather stations, both near and remote from air routes.
2.   We recommend that the radiative properties of representative contrails and contrail-produced cirrus clouds be determined.
3.   We recommend that the significance, if any, of ice crystals falling from contrail clouds as a source of freezing nuclei for lower clouds be determined.

2.6
Supersonic Transports (SSTs) in the Stratosphere

The stratosphere where SSTs will fly at 20 km (65,000 feet) is a very rarefied region with little vertical mixing. Gases and particles produced by jet exhausts may remain for 1 to 3 years before disappearing.

We have estimated the steady-state amounts of combustion products that would be introduced into the stratosphere by the Federal Aviation Agency projection of 500 SSTs operating in

1985–1990 mostly in the Northern Hemisphere, flying 7 hours a day, at 20 km (65,000 feet), at a speed of Mach 2.7, propelled by 1,700 engines like the GE-4 being developed for the Boeing 2707-300. We have used General Electric (GE) calculations of the amount of combustion products because no test measurements exist. In our calculations we used jet fuel of 0.05 percent sulfur. We have been told that a specification of 0.01 percent sulfur could be met in the future at higher cost.

We have compared the amounts that would be introduced on a steady-state basis with the natural levels of water vapor, sulfates, nitrates, hydrocarbons, and soot in the stratosphere. We have also compared these levels with the amounts of particles put into the atmosphere by the volcano eruption of Mount Agung in Bali in 1963.

Based on these calculations, we have concluded that no problems should arise from the introduction of carbon dioxide and that the reduction of ozone due to interaction with water vapor or other exhaust gases should be insignificant. Global water vapor in the stratosphere may increase 10 percent, and increases in regions of dense traffic may be 60 percent.

Very little is known about the way particles will form from SST-exhaust products. Depending upon the actual particle formation, particles from these 500 SSTs (from SO_2, hydrocarbons, and soot) could double the pre-Agung eruption global averages and peak at ten times those levels where there is dense traffic. The effects of these particles could range from a small, widespread, continuous "Agung" effect to one as big as that which followed the Agung eruption. (The analogy between the SST input and that by the Mount Agung eruption is not exact.) The temperature of the equatorial stratosphere (a belt around the earth) increased 6° to 7°C after the eruption and remained at 2° to 3°C above the pre-Agung level for several years. No apparent temperature change was found in the lower troposphere.

Clouds are known to form in the winter polar stratosphere. Two factors will increase the future likelihood of greater cloudiness in the stratosphere because of moisture added by the SSTs: the increased stratospheric cooling due to the increasing CO_2 content of the atmosphere and the closer approach to saturation indicated by the observed increase of stratospheric moisture. Such an increase in cloudiness could affect the climate. The introduction of particles into the stratosphere could also produce climatic effects by increasing temperatures in the stratosphere, with possible changes in surface temperatures.

A feeling of genuine concern has emerged from these conclusions. The projected SSTs can have a clearly measurable effect in a large region of the world and quite possibly on a global scale. We must, however, emphasize that we cannot be certain about the magnitude of the various consequences.

Recommendations

1. We recommend that uncertainties about SST contamination and its effects be resolved before large-scale operation of SSTs begins.

2. We recommend that the following program of action be initiated as soon as possible:

a. Begin now to monitor the lower stratosphere for water vapor, cloudiness, oxides of nitrogen and sulfur, hydrocarbons, and particles (including the latter's composition and size distribution).

b. Determine whether additional cloudiness or persistent contrails will occur in the stratosphere as a result of SST operations, particularly in certain cold areas, *and* the consequences of such changes.

c. Obtain better estimates of contaminant emissions, especially those leading to particles, under simulated flight conditions and under real flight conditions, at the earliest opportunity.

d. Using the data obtained in carrying out the preceding three recommendations, estimate the change in particle concentration in the stratosphere attributable to future SSTs *and* its impact on weather and climate.

3. We recommend implementation now of a special monitoring program for the lower stratosphere (about 20 km or 60,000 to 70,000 feet) to include the following activities:

a. Measurement by aircraft and balloon of the water vapor content of the lower stratosphere. The area coverage required is global, but with special emphasis on areas where it is proposed that the SST should fly.

b. Sampling by aircraft of stratospheric particles, with subsequent physical and chemical analysis.

c. Monitoring by lidar (optical radar) of optical scattering in the lower stratosphere, again with emphasis on the region in which heavy traffic is planned.

d. Monitoring of tropospheric carbon monoxide concentration because of its potential effects on the chemical composition of the lower stratosphere.

2.7

Atmospheric Oxygen: Nonproblem

Atmospheric oxygen is practically constant. It varies neither over time (since 1910) nor regionally and is always very close to 20.946 percent. Calculations show that depletion of oxygen by burning all the recoverable fossil fuels in the world would reduce it only to 20.800 percent. It should probably be measured every 10 years to be certain that it is remaining constant.

2.8

Surface Changes and the Climate

The most important properties of the earth's surface that have a bearing on climate and are likely to be affected by human activity are reflectivity, heat capacity and conductivity, availability of water and dust, aerodynamic roughness, emissivity in the infrared band, and heat released to the ground.

Since the amount of carbon dioxide in the atmosphere is dependent on the biomass of forest lands which serves as a reservoir, widespread destruction of forests could have serious climatic effects. Population growth or overgrazing that increases the arid or desert areas of the earth creates conditions that allow the introduction of dust particles to the atmosphere.

Other important surface changes are from man's activities that modify snow and ice cover, particularly in polar regions, and from some possible projects involving the production of new, very large water bodies. Increased urbanization is of possible global importance only as it produces extended areas of contiguous cities. Still, it is not certain whether effects of urbanization extend far beyond the general region occupied by the cities.

Recommendation

We recommend that before actions are taken which result in some of the very extensive surface changes described mathematical models be constructed which simulate their effects on the climate of a region or, possibly, of the earth.

2.9
Thermal Pollution

Although by the year 2000 global thermal power output may be as much as six times the present level, we do not expect it to affect global climate. Over cities it does already create "heat islands," and as these grow larger they may have regional climatic effects. We recommend that these potential effects be studied with computer models.

3.
Ecological Effects of Man's Activities

3.1

General Effects

Man produces more than a million different kinds of products, both as waste and as useful products that eventually end up as waste. We are mobilizing many materials at rates greater than the global rates of geological erosion and deposition, great enough to change their global distributions. We are using more than 40 percent of the total land surface and have reduced the total amount of organic matter in land vegetation by about one-third.

An estimate is needed for the ecological demand, a summation of all of man's demands upon the environment, such as the extraction of resources and the return of wastes. Such demand-producing activities as agriculture, mining, and industry have global annual rates of increase of 3, 5, and 7 percent, respectively. An integrated rate of increase is estimated to be between 5 and 6 percent per year, in comparison with an annual rate of population increase of only 2 percent. It is only through such a concept as ecological demand that man can assess his impact on the biosphere.

Natural ecosystems still provide us many services. Almost all potential plant pests are controlled naturally. Insects pollinate most vegetables, fruits, berries, and flowers. Commercial fish are produced almost entirely in natural ecosystems. Vegetation reduces floods, prevents erosion, and air-conditions and beautifies the landscape. Fungi and minute soil animals work jointly on plant debris and weathered rocks to produce soil. Natural ecosystems cycle matter through green plants, animals, and decomposers, thus eliminating wastes. Organisms regulate the amount of nitrates, ammonia, and methane in the environment. On a geological time scale, life regulates the amount of carbon dioxide, oxygen, and nitrogen in the atmosphere. Natural ecosystems also serve important recreational and aesthetic needs of man.

While some of these services will cease only when life is virtually annihilated, many others are easily impaired. However, these losses are gradual and progressive without discrete steps of change. The gradual attrition of natural systems results from most types of environmental pollution and thus measures the total impact of man upon his environment.

The health and vigor of ecological systems are easily reduced if (1) general and widespread damage occurs to the predators, (2) substantial numbers of species are lost, or (3) general biological activity is depressed. Most pollutants that affect life have some effect on all three processes. To prevent further deterioration of the biosphere and to repair some of the present damage, effective environmental management systems are urgently needed.

Recommendations

1. We recommend an intensive program of technology assessment. Research on the toxic effects of pollutants needs to be greatly expanded, especially to include the difficult experiments that are based on low levels of chronic exposure. We also need to have much better knowledge of the current sources of pollution, their kinds and rates, as well as projections of future trends. Both of these information needs should be part of continuing studies of the impact of technology that are closely integrated into the time phasing of planned technological development.

2. We recommend a systematic program of environmental assessment. We need more information on the routes of distribution of pollutants, their eventual distribution in the environment, and their passage through ecosystems. The present disorganized system leads to faddism and thus to the development of information on one pollutant with the neglect of others and develops no regular assessment of trends through time. Specifically, we recommend the following:

a. Early establishment of ecological base-line stations in remote areas that would provide both specific monitoring of the effects of known problems and warnings of unsuspected effects.

b. Central coordination and, where necessary, modification of national and regional surveys of critical populations of fish, birds, and mammals from commercial catches, harvests, and surveys. This would provide an early warning system by monitoring highly sensitive and vulnerable species.

c. Implementation of a number of simple measures to determine the present states of ecosystems. Collected systematically the following information would be of great value: the rates of recruitment (the reproduction and survival of young to maturity) of populations of birds and fish; the area damage to leaves of trees; the degree of oxygen depletion in deep water; and the diversity of species collected in plankton nets, soil samples, and insect light traps. All of these are indicators of ecosystem function.

d. Implementation of a 1,000-sample base-line survey of the oceans to provide general knowledge of the distribution of man-made products in the oceans. The results of such a survey would make it possible to specify the volume and distribution of observations necessary to monitor critical environmental problems in the oceans.

e. Examination, either as part of the ocean base-line survey or independently, of glaciers and sediments to help remedy the current lack of adequate historical record of the oceans and of world climate and, especially, to clarify at least the recent variations of atmospheric and oceanic particulate content.

3.   We recommend a comprehensive program of problem evaluation. Existing and emerging environmental problems must be analyzed in the broader context of social, economic, and political problems. We need think centers devoted to conflict resolution between man and environment. Substantive issues include growth in population, growth in ecological demand, a new land ethic, achieving early action in high-risk situations, allocating costs to promote better technical solutions, and obtaining effective management in international waters and airsheds. Analysis should include value changes in the traditional rights and goals of individuals, industry, and government. Fundamental changes in life style should be identified that will permit us to develop a system in which freedom from the constraints of nature is compatible with the continuing function of ecological systems.

4.   We recommend an extensive program of public education. The results of the programs here recommended must be presented to the public in a simple and understandable form through educational institutions and the news media.

3.2
DDT and Related Persistent Pesticides

Pesticides can have widespread ecological effects. The use of pesticides on crops generally requires continued and increased use of different and stronger pesticides. This is the result of a complex ecological system in which the reduction of one pest and several innocuous (to man) predators allows new pests to become dominant.

DDT can also have specific effects on species other than pests. For example, the eggshells of many birds are becoming thinner, reducing hatching success. In several species these effects now seriously threaten reproductive capabilities. Damage to these predators in an ecological system tends to create a situation in which pest outbreaks are likely to occur.

The concentrations and effects of DDT in the open oceans are not known. There are no reliable estimates and no direct measurements have been made. It is known that large amounts leave the area of application through the atmosphere and are transmitted through the world, and some portion of this falls into the oceans.

DDT collects in marine organisms. Detrimental effects have not been observed in the open ocean, but DDT residues in mackerel caught off of California have already exceeded permissible tolerance levels for human consumption. It is known that reproduction of fresh-water game fish are being threatened.

The effect of DDT on the ability of ocean phytoplankton to convert carbon dioxide into oxygen is not considered significant. The DDT concentration necessary to induce significant inhibition exceeds expected concentrations in the open ocean by ten times its solubility (1 part per billion) in water.

Recommendations

1.   We recommend a drastic reduction in the use of DDT as soon as possible *and* that subsidies be furnished to developing countries to enable them to afford to use nonpersistent but more expensive pesticides as well as other pest control techniques.

2.   In order to obtain information about the concentrations and effects of DDT in the marine environment, a base-line pro-

gram of measurement should be initiated (also recommended in the previous section). This might involve taking about 1,000 samples at selected locations and analyzing them over the course of a year. A full-scale monitoring program should await the results of such a program.

3.   We recommend greatly increased effort and support for the research and development of integrated pest control, combining a minimal use of pesticides with maximal use of biological control.

3.3
Mercury and Other Toxic Heavy Metals

Many heavy metals are highly toxic to specific life stages of a variety of organisms, especially shellfish. Most are concentrated in terrestrial and marine organisms by factors ranging from a few hundred to several hundred thousand times the concentrations in the surrounding environment.

The major sources of mercury are industrial processes and biocides. The former are often introduced into waters through municipal sewage systems. Although the use of mercury in pesticides is relatively small, it is a direct input into the environment.

Recommendations

1.   We recommend that all pesticidal and biocidal uses of mercury be drastically curtailed, particularly where safer, less-persistent substitutes can be used.

2.   We recommend that data be obtained on the concentrations of mercury in selected organisms and on its effect on ecosystems.

3.   We recommend that all industrial wastes and emissions of mercury be controlled and recovered to the greatest extent possible.

4.   We recommend that world production and consumption figures for mercury be obtained.

3.4
Oil in the Ocean

It is likely that approximately 2 million tons of oil are introduced into the oceans every year through ocean shipping, offshore drilling, and accidents. Very little is known about the effects of oil in the oceans on marine life. Present results are conflicting. The effects of one oil spill that has been carefully observed indicates severe damage to marine organisms. Observations of other spills have not shown such a marked degree of damage. Different kinds of damage have been observed for different spills.

Potential effects include direct kill of organisms through coating, asphyxiation, or contact poisoning; direct kill through exposure to the water-soluble toxic components of oil; destruction of the food sources of organisms; incorporation of sublethal amounts of oil and oil products into organisms, resulting in reduced resistance to infection and other stresses or in reproductive failures.

Recommendations
1.   We recommend that much more extensive research be undertaken to determine the action of oil in the ocean and its effects on marine biota. Future oil spills should be systematically studied beginning immediately after they occur so that a comprehensive analysis of the effects can be developed over time. Sites of previous spills should be reexamined to study the effects in sediments.
2.   We recommend that political and legal possibilities be explored for the establishment of more effective international control measures for oil-carrying tankers.
3.   We recommend that the possibility of recycling used oil be explored.

3.5
Nutrients in Coastal Waters
Eutrophication of waters through overfertilization (principally with phosphorus and nitrogen) produces an excess of organic matter that decomposes, removing oxygen and killing the fish. Estuaries are increasingly being eutrophied. Pollution of coastal regions eliminates the nursery grounds of fish, including many commercial species that inhabit the oceans.

Approximately 60 percent of the phosphorus causing overenrichment of water bodies comes from municipal wastes. Urban and rural land runoff contribute the remainder. A major contributor to the latter is runoff from feedlots, manured lands, and eroding soil.

Trends in both nutrient use and loss are rising. Fertilizer consumption is expected to increase greatly in both developed and developing countries in the next decade, increasing the nutrient runoff from agricultural lands. Concentration of animal production will continue, with the result that losses of nutrients from feedlot runoff will rise sharply. Urban waste production is also expected to increase rapidly, resulting in greater potential loss of nutrients directly into coastal waters.

Recommendations

1.   We recommend that technology be developed to reclaim and recycle nutrients in areas of high concentrations, such as sewage treatment plants and feedlots.

2.   We recommend that the dumping of industrial wastes into sewage systems be restricted so that toxic wastes do not interfere with nutrient recovery and recycling.

3.   We recommend that the use of nutrients in materials that are discharged in large quantities into water or air be avoided. For example, phosphates in detergents should be replaced with new materials, being certain that the substitute does not itself create a new problem.

4.   We recommend that the institutional structures responsible for defining, monitoring, and maintaining water-quality standards over large areas be improved. The multiplicity of authorities involved in river basins, estuaries, and coastal oceans makes effective control nearly impossible.

3.6

Wastes from Nuclear Energy

In our selection of problems for intensive study, it was necessary to omit some areas of great importance. One of these areas is that of perpetual management of large quantities of radioactive wastes which are by-products of nuclear power generation. No other environmental pollutant has been so carefully monitored and contained. This class of pollutants will, however, grow significantly in quantity over the next several decades. Therefore, it is important that attention be focused on any environmental problems that might arise.

Recommendation

We recommend that an independent, intensive, multidisciplinary study be made of the trade-offs in national energy policy between fossil fuel and nuclear sources, with a special focus on problems of safe management of the radioactive by-products of nuclear energy, leading to recommendations concerning the content and scale and urgency of needed programs.

4.
Implications of Change and Remedial Action

4.1

Introduction

The expansion and refinement of our knowledge and understanding are the necessary conditions for effective change in the present state of environmental management. However, these are not sufficient conditions. Even after optimal improvements have been made in our knowledge concerning the nature of key pollutants, their effects, their sources, their rates of accumulation, the routes along which they travel, and their final reservoirs, the questions will remain of how to apply our knowledge constructively and how to cope with the collateral consequences. As a practical matter, questions of environmental management will have to be faced before we have all the appropriate scientific and technical data, and this further complicates efforts of change or of remedial action.

In examining a wide range of specific problems at this Study, we have identified several aspects that are common to most of them and to many other critical environmental problems. These implications of change and remedial action are briefly discussed now.

4.2

Establishing New Priorities

Earlier in our history, the prevailing value system assigned an overriding priority to the first-order effects of applied science and technology: the goods and services produced. We took the side effects—pollution—in stride. A shift in values appears to be under way that assigns a much higher priority than before to the control of the side effects. This does not necessarily imply a reduced interest in production and consumption. When the implications of remedial action and the choices that must be made become clear, there may be second thoughts, confusion, and feelings of frustration.

In the effort to arrive at an optimal balance in specific situations, something will have to give. But the old routine assumption that it is the environment that must give has become intolerable. This assumption must be rejected in favor of an optimal balance to be reached from a point of departure in affixing the responsibilities for pollution.

4.3
Affixing Responsibilities

As a point of departure for taking action, we recommend a principle of presumptive "source" responsibility. While remedial measures can be attempted on the routes along which pollutants spread or in the reservoirs in which they accumulate, we believe that these measures should be generally taken at the "sources," which we define broadly to include (1) sources or the points in the processes of production, distribution, and consumption, at which the pollutant is generated, for example, factories, power plants, stockyards, bus lines; (2) protosources or earlier points that set the conditions leading to the emission of pollutants at a later stage, for example, the manufacturers of automobiles that emit pollutants when driven by motorists, or the brewers of beer sold in nonreturnable cans that are tossed aside by the consumer; and (3) secondary sources or points along the routes where pollutants are concentrated before moving on to the reservoirs, for example, sewage treatment plants or solid waste disposal centers.

The principle does not connote any element of blame or censure, nor is it intended to foreclose a judgment concerning where the financial costs of correction should ultimately be borne. It is intended, however, to indicate a point of departure for analysis and action. It rests, in part, on the basis that, if something goes wrong, it should be traced to its origin and corrected in terms of its cause; in part on a hypothesis that the source, protosource, or secondary source will typically be in the best position to take corrective measures, whether alone or with help from others; and in part on the view that the remedies available, the criteria for choice among them, and the implications of remedial action can best be appraised at the sources as here defined.

4.4
Accepting the Costs

Remedial changes will ordinarily involve financial costs, and the costs may be large in relation to the scale of the source enterprise. If the source enterprise can neither absorb the cost nor pass it on, it will be necessary to face a choice among failure of the enterprise, continuance of the pollution, or financial assistance out of public revenues. The initial change may have consequences reaching past the source enterprise to its employees, its suppliers, and its customers and beyond in widening waves of change that may engulf deep-rooted patterns of economic and social behavior. Our society is familiar with far-reaching readjustments caused by technological innovation or organizational change in the past. Comparable readjustments may be required by changes instituted to control pollution.

4.5
Assessing the Available Means for Action

The means available within the political process and legal system to encompass remedial changes include taxes designed as incentives, stimuli, or pressures, regulations, typically involving a statute, an administrative agency, and supplementary action through the courts; common-law remedies in the courts, incrementally adjusted to contemporary needs; governmental financing of research and assistance to facilitate costly adjustments to desired changes; and governmental operations, civilian and military. Governmental action in its own house can have a dual importance: in itself and as a model for others to follow.

4.6
Stimulating Effective Actions

The political, legal, and market processes of our society are profoundly affected by the nature and quantity of information available and the manner in which the information is infused into them. It is neither necessary nor feasible to postpone recommendations for action until scientific certainty can be achieved. The political process is accustomed to decisions in the face of uncertainty on the basis of a preponderance of the evidence or substantial probabilities or a reasonable consensus of informed judgment.

Thus, it is not enough for scientists and technologists to expand and refine their knowledge. They must also present their knowledge in a manner that clearly differentiates fact, assertion, and opinion and facilitates the task of relating the data to the possibilities of corrective action. But if such information is to be used, the Congress and state legislative bodies must be provided with instrumentalities and qualified staff to enable them more effectively to sort out and utilize the input of data, proposals, complaints, and suggestions that will flow into them in increasing volumes from all sectors of our society.

4.7
Developing New Professionals

In addition to general public education, we stress the special importance of some changes in scientific, technical, and professional education and training. A sensitivity to the relations between the processes of production, distribution, and consumption, on the one hand, and the processes of pollution, on the other, and a disposition to explore all the potentialities of technology and organization in the search for an optimal balance should be incorporated into their training. This applies to economists, lawyers, and social scientists as well as to scientists and engineers. Individual contributions may be undramatic now, but over time they will be critical.

4.8
Cooperating with Other Nations

Although many problems are global in nature, the solutions to these problems will generally require national as well as international action. Typically, remedial measures within one nation will need support from parallel actions within other nations. Frequently, collaborative international action will be required. The prospects for such cooperation are best for programs of collection and analysis of data. International cooperation on monitoring may also increase the likelihood of smooth relations should a global program ever demand strict international regulation or control of pollution-producing activities.

In the foreseeable future the advanced industrial societies will probably have to carry the major burden of remedial action. Developing nations are understandably concerned far more with economic growth and material progress than with second-order effects of technology. Similar attitudes were prevalent in the early stages of growth of present industrialized nations.

The challenges of international cooperation and collaboration in the critically important environmental areas studied by SCEP will be before the United Nations Conference on the Human Environment in 1972. We hope that this Report will provide useful inputs to that Conference and that the Study model furnished by SCEP will be applied to other critical problems of the environment.

Part II
Reports of SCEP Work Groups

1.
Work Group on Climatic Effects

Chairman
William W. Kellogg
NATIONAL CENTER FOR
ATMOSPHERIC RESEARCH

Richard D. Cadle
NATIONAL CENTER FOR
ATMOSPHERIC RESEARCH

Robert G. Fleagle
UNIVERSITY OF WASHINGTON

Stanley Greenfield
RAND CORPORATION

Christian E. Junge
MAX-PLANCK INSTITUT FÜR
CHEMIE

Charles D. Keeling
SCRIPPS INSTITUTION OF
OCEANOGRAPHY

Julius London
UNIVERSITY OF COLORADO

Lester Machta
ENVIRONMENTAL SCIENCE
SERVICES ADMINISTRATION

J. Murray Mitchell, Jr.
ENVIRONMENTAL SCIENCE
SERVICES ADMINISTRATION

Reginald E. Newell
MASSACHUSETTS INSTITUTE OF
TECHNOLOGY

Hans A. Panofsky
PENNSYLVANIA STATE UNIVERSITY

James T. Peterson
NATIONAL AIR POLLUTION CON-
TROL ADMINISTRATION

Walter Ramberg
DEPARTMENT OF STATE

G. D. Robinson
CENTER FOR THE ENVIRONMENT
AND MAN, INC.

Silvio G. Simplicio
WEATHER BUREAU

Joseph Smagorinsky
ENVIRONMENTAL SCIENCE SERV-
ICES ADMINISTRATION

Howard J. Taubenfeld
SOUTHERN METHODIST UNIVER-
SITY SCHOOL OF LAW

Morris Tepper
NATIONAL AERONAUTICS AND
SPACE ADMINISTRATION

F. Joachim Weyl
HUNTER COLLEGE OF THE CITY
UNIVERSITY OF NEW YORK

Rapporteur
William G. Anderson
HARVARD LAW SCHOOL

1.1
Introduction
1.1.1
Our Frame of Reference

We have chosen to restrict ourselves as far as possible to atmospheric problems that are global in scale and critical in the sense that they may affect the environment. We have emphasized those aspects where man's activities seem to have an influence on the atmosphere, but we are keenly aware that one cannot judge this impact without an understanding of the natural background or the competition. Thus, we have dealt with a wide variety of atmospheric features and their man-made alterations, but we have had to ignore some also, such as local weather modification and urban air pollution.

There is another self-imposed limitation to our frame of reference, in that we have not considered any ecological effects of atmospheric change or pollutants. Actually, we are not aware of any global effects of air pollution on living things, though of course the air is a carrier of some persistent compounds that can build up to harmful levels in the oceans. This was of great concern to the Work Group 2, which dealt with the matter of accumulation of persistent pesticides and heavy metals.

Perhaps the most substantial contribution we can make to the matter of global atmospheric changes by man is the combination of meteorological, chemical, and economic expertise. The atmospheric scientists can, for the first time in some instances, work with quantitative inputs concerning the amounts and distributions of man-made contaminants of various kinds and can weigh these against natural sources and sinks. Even though we cannot trace all the implications now, the material will be available and in a form such that we and others can continue to work with it.

We have organized our report in such a way that it deals first with the way various contaminants are introduced, carried, dispersed, and removed from the atmosphere; then we deal with changes that man may produce through his use of the earth's solid and watery surface; and finally we deal with the effects of each of these elements on the climate.

1.1.2

The Atmosphere

Planet Earth has a compressible atmosphere made up of nitrogen (N_2) and oxygen (O_2) in amounts that stay remarkably constant in time. In addition, there are a number of other minor gases that are reasonably constant atmospheric constituents. Also present, however, are gases found in relatively small and in some cases highly variable concentrations such as water vapor, carbon dioxide, and ozone, each of which plays a significant role in determining the temperature structure and the energy transfer of the atmosphere.

It is important that water can exist in all three phases (ice, liquid, vapor) in the earth-atmosphere system—and that transformations take place among all three phases. Some of these changes, such as condensation from water vapor to liquid in the formation of clouds, involve the release of latent heat and therefore represent an important physical link in the energy cycle of the atmosphere.

The amount of water vapor present in the atmosphere depends on many factors and in the lower atmosphere is highly variable. In some cases, as over tropical areas, the concentration relative to air (mixing ratio) can be as high as 0.04 by mass. Water vapor mixing ratio decreases with height to a constant value in the stratosphere of about 3×10^{-6} (3 ppm by mass).

The concentration of carbon dioxide is nearly constant in the troposphere and stratosphere at a mixing ratio of about 320 ppm by volume, whereas ozone exists in relatively small amounts in the troposphere (except for very local areas) and has its maximum concentration at about 25 km. Its mixing ratio at that altitude can be of the order of 10 ppm by volume.

In addition to these gases, the atmosphere acquires from natural and man-made sources a number of variable trace gases (such as SO_2, oxides of nitrogen) and solid particles such as dust, sea spray, and sulfates. The latter are important in the atmosphere, primarily because they provide nuclei for condensation and freezing but also because they are involved with radiative scattering and absorption processes and therefore have an influence on the temperature.

The temperature distribution in the atmosphere is the result of many interacting processes involving solar and terrestrial radiation and motions within the atmosphere itself. On the average the earth receives solar radiation in the amount of about 2 cal cm^{-2} min^{-1} at the top of the atmosphere on a unit surface perpendicular to the sun's rays (or about 0.5 cal cm^{-2} min^{-1} when spread over the entire globe. Although it is not certain, it is believed that there is very little (if any) variation in this "solar constant."

As the solar beam passes through the high atmosphere, a very small amount is absorbed. In the stratosphere, ozone absorbs about 3 percent of the incoming solar radiation, and this absorption is responsible for the relatively high temperature found in the stratospheric region from 25 to 50 km. As the solar beam penetrates the lower atmosphere, it is reduced still further by absorption (mainly in water vapor) and by back reflection (mainly from clouds, but also from air molecules and large dust-type particles). The average total absorption of solar radiation by atmospheric gases, clouds, and dust amounts to a little more than 20 percent of the incoming solar radiation.

The average reflectivity (planetary albedo) of the earth-atmosphere system is approximately 30 to 55 percent, with the clouds being responsible for over three-fourths of this amount. Thus, just under 50 percent of the incoming radiation is absorbed at the earth's surface. It is important to note, however, that there are large variations around the earth in this absorption, due to the varying nature of the underlying surface. In the polar regions, for instance, where there is ice and snow, the surface reflectivity is very high, and little of the solar energy received at the surface is absorbed. On the other hand, in the tropics, where the oceans represent a large part of the surface area, the reflectivity is relatively low, and the absorbed energy is used for the most part in evaporating water from the ocean surface.

The earth emits infrared radiation. This radiation is almost completely absorbed by the principal polyatomic gases in the atmosphere (H_2O, CO_2, and O_3), which in turn reemit the radiation both upward (eventually out to space) and downward (back to the ground). Clouds absorb and emit infrared radiation also and, in general, do so as black bodies (that is, very efficiently). The amount of radiation returned to the ground, therefore, depends principally on the water vapor content, carbon dioxide content, and cloudiness of the atmosphere. Changes in these variables would, of course, affect the radiation budget of the atmosphere.

On the average the atmosphere continuously loses energy as a net result of these radiative processes. The energy is replenished in two ways. First, there is conduction of heat from the ground to the atmosphere. Second, and more important, solar radiation absorbed at the ocean surface evaporates water, and, when this water vapor finally condenses in the atmosphere, its latent heat is released to the air. Approximately two-thirds of the energy transfer from the surface to the air is accomplished by this latter mechanism.

Since incoming solar radiation on a unit horizontal surface is more direct in equatorial than in polar regions, more than twice as much solar radiation on an annual average is available at the equator as is available at the poles. Outgoing radiation depends strongly on the temperature of the radiating substance, but, since the temperature in the atmosphere does not vary much (on an absolute scale) from equator to pole, there is only a slight decrease with latitude of emitted radiation. As a result there is an excess of incoming solar radiation over emitted infrared radiation in equatorial regions and a deficit in polar regions. If the earth did not rotate, this would be somewhat analogous to heating a fish tank at one end and cooling it at the other. In the fish tank this would give rise to a direct circulation of the fluid—rising at the heated end, flowing toward the cooled end, descending and returning to the heated end.

The large-scale circulation of the real atmosphere, however, although forced initially in response to the unequal geographic heating and cooling, is much more complicated, due primarily to the effect of the earth's rotation. There are many additional perturbing factors affecting the circulation, among which are the uneven distribution of land and water and the roughness of the ground.

One obvious consequence of large-scale motions for the general structure of the atmosphere is that the average latitudinal variation of temperature that would obtain, if the temperature distribution resulted solely from a balance of radiative factors, is decreased. Atmospheric motions are also instrumental in adjusting the average vertical temperature distribution and, together with radiative processes, help to produce the different characteristics of the troposphere and stratosphere. The qualitative difference in lapse rate (vertical temperature change) between troposphere (decreasing temperature with height) and stratosphere (constant or increasing temperature with height) and the action of scavenging processes by rain and snow lead to an important difference in the residence times of various atmospheric constituents in these two regions. Gases and particles in the troposphere are fairly well mixed vertically within periods of a few days to about a month and at the same time are removed by scavenging and by direct contact with the surface. In the lower part of the stratosphere, however, their residence times are of the order of two years.

The net result of the various highly interactive processes and features of the atmosphere is a long-term average weather and climate pattern which has some general features that are reasonably well understood, for example, the mean latitudinal and vertical temperature distribution, the existence of trade winds in the ocean areas of the tropics, and so forth. Many details of these patterns, however, involving both subtle and very obvious variations from year to year in circulation, temperature, and rainfall, cannot be satisfactorily explained at present. These irregular variations exist even with the basic features of the earth-atmosphere system and of the sun unchanged. But if some of these features were to change, the statistical properties of the atmosphere would probably undergo additional variations.

We know that in the earth's history the climate has changed on many time scales, and many different theories have been invoked to explain these changes. These theories involve variations of the sun's radiation, the earth's orbit, the earth's surface, the composition of the atmosphere, or the circulation of the oceans (Mitchell, 1965, 1968; Sellers, 1965). Some of these variations occur naturally, as, for example, changes in the earth's orbital elements (which can be calculated and forecast quite precisely) and changes in the number of particles due to volcanic eruptions (which cannot at present be forecast). Other variations, so far fairly modest, are the result of man's activity, for example, the recent change in the concentration of carbon dioxide. One thing appears clear: climate has changed over decades, centuries, and millennia without the influence of human activity.

Between 1900 and approximately 1945, global temperatures rose at a rate of about 0.8°C per century, and glaciers receded. Since then a cooling trend has set in, at a somewhat slower rate than the previous warming. There is a tendency to blame these changes on some effect of increasing population and industrialization, but the magnitudes of those climatic changes are no larger than earlier changes that cannot be attributed to man. This does not mean that man's activities may not have influenced climatic changes during the twentieth century; merely that we do not know how the climate would have changed if man had not been present. Man-made climatic changes could not have been larger than natural changes—and they may have been a good deal smaller.

Still, we suspect that this situation may not last; eventually man can become so numerous and powerful that his activities will change the composition of the atmosphere or the character of the earth's surface enough to produce effects larger than those "naturally" expected. Ice advances and retreats, widespread droughts, changes of the ocean level, and so forth, were accompanied by only slight shifts in the mean circulation pattern and only small changes in the average temperature over large parts of the earth. Year-to-year variations in the rainfall, temperature, and circulation characteristics of the atmosphere are much larger than the mean atmospheric properties that are found when the climate changes. It is likely that there exists in the atmosphere an almost continuous set of possible climatic regimes; that a small change in a mean condition of the atmosphere may be accomplished by a relatively small perturbation in one or more of the relevant parameters; and that small changes in the mean conditions of the atmosphere or the surface can then produce a change from one climatic regime to another (see, for example, Lorenz, 1970).

This thought will emerge repeatedly in the sections that follow, as we try to assess the influence of man on the climate. We will be seeking to identify those "leverage points" that man can reach, points where his relatively subtle alterations of the environment could influence significantly the global climate. It is in the interest of a rational society to be on the lookout for any such changes and to develop theories of atmospheric behavior sufficient to allow us to forecast the atmosphere's future course, given a knowledge of what man will be doing. The effort expended will certainly be trivial compared to the possible return (Hess, 1959; Lamb, 1966; Fletcher, 1969; Flohn, 1969).

1.2
Man-Made Atmospheric Contaminants

When we consider the ways that man can modify the envelope of air that covers his planet, his introduction to it of contaminants in ever greater amounts comes to mind first. Therefore, our first purpose will be to point to the magnitudes of the changes in the atmosphere due to man's contaminants (including waste heat) and to say as much as we can about trends and future levels of these contaminants.

1.2.1
Carbon Dioxide and Other Trace Gases That May Affect Climate

Carbon Dioxide

ATMOSPHERIC CONTENT OF CO_2. Carbon dioxide (CO_2) is a trace gas in the earth's atmosphere with a concentration of a little over 0.03 percent by volume (about 320 ppm). In spite of its relatively small concentration it plays an important role in determining the temperature of the planet. It absorbs sunlight to a modest degree; more importantly, it absorbs (and emits) infrared radiation. By intercepting a part of the infrared radiation that is emitted by the earth's surface and reradiating it back toward the earth, it cuts down the rate of surface cooling, and at the same time it acts to cool the upper atmosphere (see Section 1.4.2).

It is this influence on the heat balance of the earth-atmosphere system that arouses our concern over any change in its concentration. The idea that climatic change could result from changes in atmospheric CO_2 content was suggested independently by the American geologist, T. C. Chamberlain, in 1899, and the Swedish chemist, S. Arrhenius, in 1903. It was rather easy to show that the increasing production of CO_2 by the burning of fossil fuels (coal, oil, and natural gas) *ought* to influence the amount of

CO_2 in the air. Nevertheless, it was not until the period of the International Geophysical Year in 1958 that the first systematic and accurate observations of its concentration were begun.

There are now observations, some for only limited periods, from Swedish aircraft (Bolin and Bischof, 1969), Point Barrow (Kelley, 1969), the Antarctic (Brown and Keeling, 1965), and Mauna Loa (Pales and Keeling, 1965), the latter being the most continuous and frequent. Figure 1.1 shows the trends of yearly mean values of CO_2 concentration at these four places, and Figure 1.2 shows the changes in monthly mean values for Mauna Loa.

From these observations the following generalizations can be made:

1. CO_2 seems to have been increasing throughout the world at about 0.2 percent per year, or 0.7 ppm out of 320 ppm. The dashed "best fit" curve in Figure 1.2 shows that this rate is not constant, but the length of record is too short to place much emphasis on the deviations from a linear trend.

Figure 1.1   Annual Mean Values of CO_2

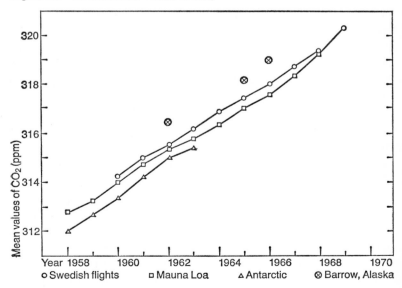

Sources: Swedish flights (Bolin and Bischof, 1969); Mauna Loa (Pales and Keeling, 1965, Bainbridge, 1970); Antarctic (Brown and Keeling, 1965); Barrow, Alaska (Kelley, 1969)

Figure 1.2 CO₂ Concentration from Burning of Fossil Fuels

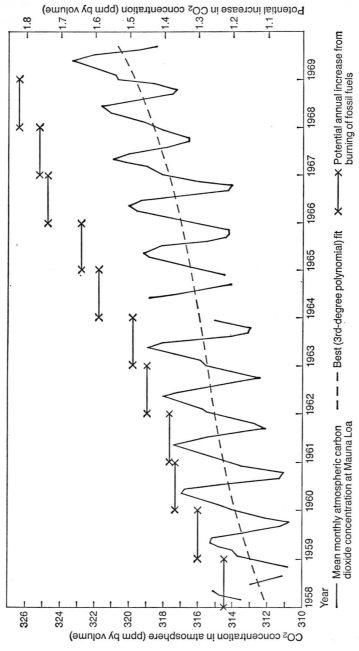

Sources: Monthly (Pales and Keeling, 1965), (Bainbridge, 1970); Best fit (Cotten, 1970); Annual increase (United Nations, *World Energy Supplies*)

2.   There is somewhat more CO_2 at the most northerly station and the least is at the Antarctic station, but the total difference is only about 1 ppm, corresponding to a lag from north polar regions to south polar regions of about 18 months.

3.   In the Northern Hemisphere there is a distinct seasonal oscillation in CO_2, with an amplitude at the surface at high latitudes in the Northern Hemisphere of about 9 ppm and at Mauna Loa (altitudes 3,398 meters, latitude 19°34′N) of 6 or 7 ppm. This oscillation decreases in amplitude with increasing altitude, according to the Swedish observations with aircraft.

4.   It is estimated that, in the interval 1850–1950, fossil fuel consumption produced an amount of CO_2 equal to 10 percent of the amount estimated to be in the atmosphere in 1950 (United Nations, 1956). Analysis of tree rings and other biological specimens of known age show that there was a 1 to 2 percent decrease of radioactive carbon-14 (a 1 to 2 percent "Suess effect") over the same period (Suess, 1965).

These are the main facts regarding the changing atmospheric CO_2 content. What are some of the related facts that help to explain this change that has been taking place? And what is our best guess about the future? The picture is far from clear, but we do have a fairly good idea of what is happening—even if we sometimes have difficulty assigning the right numbers.

In 1950, the concentration of CO_2 in the atmosphere was probably a little over 300 parts per million (ppm) by volume, or 445 ppm by mass. (It was not measured at that time, so this is an estimate.) Since the mass of the entire atmosphere is 5.14×10^{21} grams, there were about 2.3×10^{18} grams of CO_2 in the atmosphere. In that year about 0.67×10^{16} grams of CO_2 were released into the atmosphere by burning fossil fuel, and this was 0.28 percent of the amount already in the atmosphere in 1950 (President's Science Advisory Committee [PSAC], 1965). The amounts added each year have been increasing ever since, and are expected to rise further. The amounts are shown in abbreviated form in Table 1.1.

RESERVOIRS FOR CO_2. In order to predict the future CO_2 concentration, we could, as some have done, assume that the fraction of CO_2 produced by combustion of fossil fuels that has stayed in the atmosphere in the past decade (about 50 percent) will remain

Table 1.1
Carbon Dioxide Production and Accumulation

Year	Amount Added from Fossil Fuel (Mt/yr)[a]	Cumulative Amount Added over Previous Decade[a]	Concentration by Volume (ppm)[b]	Total Amount in Atmosphere (Mt)[c]
1950	6,700	52,200	306	2.39×10^6
1960	10,800	82,400	313	2.44×10^6

Note: The percentage of CO_2 remaining in the atmosphere of that introduced in 1960 can be calculated from the data in this table and by noting from Figure 1.1 that the increase of CO_2 concentration by volume was approximately 0.7 ppm. This corresponds to 5,450 Mt increase in 1960 (0.7 times 2.44×10^6 divided by 313). The first column notes that 10,800 Mt were introduced in 1960, thus the percentage increase was 51.

[a] Source: President's Science Advisory Committee (PSAC), 1965.
[b] The 1960 number is the average of the observed values in Figure 1.1. The 1950 value was obtained by a linear extrapolation of the curves in Figure 1.1.
[c] This column is computed from data in the preceding column by taking the mass of the atmosphere to be 5.14×10^{21} grams. Note that 1 mole of air weighs 29 grams.

the same in the future. This is certainly a rational first assumption, but we have the uneasy feeling that the real world is not likely to remain so simple and linear. The other 50 percent had to go into some other reservoirs, and this leads us to inquire about them and how they are likely to behave.

The two reservoirs that must be involved are, first, the oceans, and second, what we will call the biosphere, the mass of living things and nonliving organic matter on land and in the oceans, of which the forests represent the largest portion by far.

It is difficult to estimate the total mass of living things and the organic material on which they feed. Attempts to do this have arrived at numbers that are roughly one to two times the amount of carbon in atmospheric CO_2 (2.5×10^6 megatons [Mt] of CO_2 or 0.68×10^6 Mt of carbon). The biosphere would respond to an increase in atmospheric CO_2 by growing faster, since a 10 percent increase in CO_2 corresponds to an increase in the rate of photosynthesis (if living things were not nutrient or water limited) of 5 to 8 percent (Keeling, 1970). At the same time, man, by cutting down forests and removing the wood from the carbon cycle, can decrease the mass of the biosphere. Actually, however, we do not know quantitatively how the biosphere responds to a change in atmospheric CO_2 or to man's temporary removal of part of it. (This is treated further in the appendix to the report of Work Group 2.)

Turning to the oceans, it must be made clear at this point that, when we speak of the ocean reservoir, we are referring to that part of the ocean that can exchange CO_2 with the atmosphere and remain more or less in equilibrium with the atmosphere, in the course of a few years. This must be the surface layer that mixes fairly rapidly. The total ocean contains nearly 60 times as much CO_2 as the atmosphere, but the time needed to exchange that part of the water below 1 km with the surface water is probably 500 to 1,000 years, perhaps longer (Keeling, 1970). Thus, except for a few limited parts of the ocean, such as a limited portion of the North Atlantic and the Weddell Sea, where there can be an overturning of the deep waters, we are only concerned with the first few hundred meters of near surface water.

There are two kinds of oceanic variations that could have an effect on atmospheric CO_2. The overturning of the deep waters just referred to, which would bring CO_2-rich water to the surface, could release large amounts of CO_2 in a short time. Furthermore, a change of the surface water temperature over a large area would influence the CO_2 balance between ocean and atmosphere, because an increase in water temperature lowers the ability of water to hold CO_2. Therefore, though we have not taken these factors into account, changes in the ocean could have an important bearing on future CO_2 concentrations. (See report of Work Group 3 for further discussion.)

The inorganic carbon in ocean water is mostly in the form of carbonate and bicarbonate, balanced against the available metallic cations, but about 1 percent is in solution as dissolved CO_2. Taking into account the thermodynamic relationships between the chemical species, we can calculate that, although a layer of ocean water 60 m deep contains as much inorganic carbon as the atmosphere, the layer will take up only one-tenth as much CO_2 as the atmosphere will if a process such as fossil fuel combustion injects more CO_2 into the air. Based on evidence from the rate of mixing of radioactive tracers from nuclear tests, we expect between 200 and 400 m of ocean water to mix with surface water in a period of a few years, but our knowledge of this kind of mixing is very sketchy indeed, and the process must vary widely in different parts of the ocean.

This limited ability of the oceans to take up additional CO_2 in a short time, and the apparent inability of the biosphere to increase its uptake by more than a few percent, accounts for the fact that about half of the added CO_2 from fossil fuels has remained in the atmosphere (see notes on Table 1.1 and Table 1.2). This does not tell us what the ultimate capacity of the ocean and biosphere reservoirs is, but merely that together they respond by taking up about half of each new injection into the atmosphere.

In order to estimate the size of these reservoirs, it is necessary to consider the Suess effect. In general terms this effect is the percent dilution of natural carbon-14 caused by the release (as CO_2) of carbon-12 and carbon-13 to the atmosphere from the combustion of fossil fuels. Fossil fuels are very low in radioactive carbon-14, since the half-life of carbon-14 is about 6,000 years. As we have pointed out, during the century prior to 1950 the estimated additional CO_2 available to the atmosphere (based on fossil fuel usage) was 10 percent of the amount present in the atmosphere in 1950, and the observed decrease in carbon-14 was 1 to 2 percent. Taking into account that the ocean and biosphere reservoirs, plus the atmosphere, share in the process of diluting the carbon-14, the conclusion is reached that a 1 percent Suess effect implies that the total reservoir (ocean, biosphere, and atmosphere) is 10 percent/1 percent or 10 times larger than the atmospheric content alone and that a 2 percent Suess effect implies a total reservoir that is 5 times larger. This gives us an indication of the size of the ocean-plus-biosphere reservoir for $C^{14}O_2$, relative to the atmosphere, but does not determine the "sizes" of the ocean and the biosphere relative to each other.

We must emphasize that this conclusion only describes the size of the reservoir for $C^{14}O_2$ and that our concern is with the size of the reservoir for atmospheric CO_2 (of which $C^{14}O_2$ is but a very small fraction). Assuming that the portion of the biosphere that exchanges $C^{14}O_2$ in a few years time is about the same size as the atmospheric reservoir (recall the estimates of the mass of the biosphere in terms of the mass of atmospheric carbon), the biosphere plus atmosphere would represent twice the atmospheric reservoir, and the size of the oceanic reservoir, as deduced for a 2 to 1 percent Suess effect, would be three to eight times larger, respectively. Such an ocean reservoir would be equivalent to a layer of ocean water 180 to 480 m in average depth. This agrees with the statement made earlier that 60 m of ocean surface water will take up one-tenth as much injected CO_2 as will the atmosphere. We may conclude that the accessible ocean reservoir for atmospheric CO_2 is 0.3 to 0.8 times that of the atmosphere itself. The previous argument about mixing times in the surface of the ocean suggests that the smaller depth would be more likely but does not prove conclusively that the larger depth is unreasonable. FUTURE TRENDS. If the ocean reservoir is dominant, then it will take up its fraction of the added CO_2 in a regular way, in proportion to the partial pressure of atmospheric CO_2, and it will store it indefinitely. Further, since the mixed layer is slowly exchanged with the deep ocean water over a period of decades, the ocean will take up still more. If, on the other hand, the land plants of the biosphere are the main reservoir, we should expect to see the CO_2 partly returned to the atmosphere within a few decades, as the previously added growth of trees falls and begins to rot. The biosphere can take up the increased CO_2, but only temporarily. The trend toward depleting the major remaining stands of virgin forests, such as those in tropical Brazil, Indonesia, and the Congo, will further reduce this possible reservoir and may release even more CO_2 to the atmosphere.

A model has been devised by Machta, Machta, and Olson, at this Study, that takes into account the best estimate of the ocean and biosphere capacities to take up additional CO_2. The model, which resembles others that have been used in this area, was used to estimate the buildup of CO_2 until the year 2000. Its results are shown in Table 1.2.

One further point should be made concerning the longer view (beyond the year 2000). If by the year 2000 we have consumed from 2 to 12 percent of the total recoverable fossil fuel reserves of the globe and the remainder were burned, then 8 to 50 times the cumulative amount released before the year 2000 would go into the atmosphere. If half of that remained in the atmosphere, it would be within man's power in the next century

Table 1.2
Possible Atmospheric Carbon Dioxide Concentrations[a]

Year	Amount Added from Fossil Fuel (Mt/yr)[b]	Cumulative Amount Added over Previous Decade (Mt)	Concentration by Volume (ppm)	Total Amount in Atmosphere (Mt)	Percentage of Annual Addition Remaining in Atmosphere
1970	15,400	126,500	321	2.50×10^6	52
1980	22,800	185,000	334	2.61×10^6	52
1990	32,200	268,000	353	2.75×10^6	52
2000	45,500	378,000	379	2.95×10^6	51

Note: This table is not a projection or a prediction. These calculations have been developed to provide insight into the nature of problems that may exist over the next several decades.

[a] The data in this table are taken from a model of the carbon cycle developed at the Study by Jon Machta, Lester Machta, and Jerry Olson, which includes biospheric and oceanic uptake of CO_2 (see Appendix to the report of Work Group 2). The model calculates yearly values for atmospheric CO_2, using past fossil fuel CO_2 production data from the United Nations. It was assumed that the atmosphere and the oceans were in equilibrium in 1860 with a concentration of 290 ppm by volume. The mechanisms used for oceanic uptake were gaseous concentration difference and the chemical mixing of CO_2 into a larger pool of carbonates. The mechanism used for continental biospheric uptake was CO_2 fertilization, that is, an increase in atmospheric CO_2 concentration, causing a proportional increase in photosynthetic rate. According to Keeling (1970) the mass of CO_2 in that layer of the ocean which is in communication with the atmosphere is between 5 and 8 times the mass of CO_2 in the atmosphere. The model assumed a mixing layer in the ocean that contains 8 times the mass of CO_2 that is in the atmosphere. The fertilization coefficient was adjusted to agree roughly with the Mauna Loa measurements (See Figure 1.2).

[b] For illustrative purposes only, the numbers in the first column are based on 4 percent annual growth rate until 1980 and 3.5 percent thereafter. These numbers are slightly larger than those compiled by Work Group 7.

Note: One million metric tons equals one megaton (Mt) or 10^{12} grams (g).

to increase the CO_2 of the atmosphere by a factor of 4 or more. The fact that it is possible leads us to urge that its implications be investigated (see Section 1.4.2).

Other Trace Gases

The list of trace gases that exist in the atmosphere is very long and includes ozone and water vapor (which man does not produce enough of to compete with nature globally); methane, carbon monoxide, formaldehyde, and a host of other organic compounds (which do not have a significant effect on the atmospheric heat balance by themselves); and SO_2, oxides of nitrogen, and hydrocarbons (which do not have a significant effect on radiation by themselves, but which can form particles that do). We do not consider these gases to have a global significance except in so far as they form particles (see Section 1.2.2).

Conclusions

1. The atmospheric CO_2 concentration has increased at the average rate of 0.2 percent per year during the period 1958–1969.
2. For this period, approximately one-half of the input has remained in the atmosphere. The other portion has gone into the ocean and the biosphere. The partitioning of this other portion is uncertain.
3. Predictions of future atmospheric CO_2 concentrations depend both on the growth rate of fossil fuel combustion and on the partitioning of man-made CO_2 among the atmospheric, oceanic, and biospheric reservoirs. This partitioning is too uncertain to allow accurate predictions. However, the assumption of about a 4 percent annual growth in fossil fuel combustion, and of a continuation of the partitioning found during the 1958–1969 period, leads to an increase in atmospheric CO_2 over present levels of almost 20 percent in the year 2000, from 320 to 379 ppm.

Recommendation

1. We recommend that there be developed a station net with continuous analysis and with sufficiently high accuracy and resolution to determine the global trend and to analyze the partitioning of CO_2 uptake between the ocean and the biosphere (see report of Work Group 3).

In order to obtain the full benefit of this station net, supporting programs are needed, especially addressed to the gathering of pertinent oceanographic data, the procurement of stratospheric observations of trends of CO_2 concentration, and the supplementation of station data by satellite and aircraft measurements to improve geographic and vertical coverage.

1.2.2

Particles and Turbidity

Atmospheric Content of Particles

The visible effect of particles in the air (or aerosols—we shall use these terms interchangeably) is to scatter sunlight. Air molecules also scatter sunlight, but they are much smaller than the wavelengths of visible light and scatter light in a predictable and isotropic way (Rayleigh scattering). Particles of more than about 0.1-micron (μ) radius scatter in a complicated manner, with a preponderance of scattered light in the forward direction (Mie scattering) and a relatively weak wavelength dependence. They can also absorb sunlight to some extent.

The solar beam, on its way to the earth's surface, is attenuated by absorption in ozone and other trace gases, is scattered isotropically by the air molecules, and is also scattered and absorbed by aerosols. This effect can be summarized by the formula for the intensity of the solar beam at a given wavelength:

$$I = I_0 e^{-(\alpha_a + \alpha_s + \alpha_d)m}$$

Where I_0 is the intensity outside the atmosphere and where α_a is the total absorption coefficient by gases, α_s is the gaseous scattering coefficient, α_d is the "turbidity coefficient" and accounts for aerosol scattering and absorption, and m is the path length relative to the vertical path length through the atmosphere (the cosecant of the solar zenith angle). When I is measured, the turbidity coefficient, α_d, can be calculated, since the other parameters are fairly well known. Measurement on cloudless days by standardized methods of such a turbidity index is probably the simplest and soundest way to obtain a continuous record, in relative terms, of the particulate load of the atmosphere.

Unfortunately, this method of turbidity determination lumps together the loss from the beam due to both scattering and absorption by particles and makes it difficult to assess the effect of particles on the radiation balance of the earth (Bryson, 1968).

Sources

Particles in the troposphere can be most generally characterized by their sources:

1.   Natural continental aerosols:

From dust storms and desert areas, with size range above about $0.3\text{-}\mu$ radius.

From photochemical gas reactions between ozone and hydrocarbons from plants, resulting in very small particles of less than $0.2\text{-}\mu$ radius.

From photochemical reactions between trace gases such as SO_2, H_2S, NH_3, and O_3 or atomic O. Such reactions are strongly influenced by humidity or the presence of cloud droplets.

From volcanic eruptions, which emit particles of all sizes and trace gases (especially SO_2) that subsequently can become particles in the stratosphere.

2.   Natural oceanic aerosols:

From the evaporation of ocean spray. These have essentially the same composition as sea salt, with size range above about $0.3\text{-}\mu$ radius.

3.   Man-made aerosols:

From "smoke," that is, solid particles formed by combustion.

From photochemical gas reactions between unburned or partially burned organic fuel (for example, gasoline) and oxides of nitrogen ("smog"). This results in small particles, initially, of less than $0.2\text{-}\mu$ radius.

From photochemical reactions between SO_2 and O_3 or atomic oxygen, such reactions being essentially the same as those of natural SO_2.

While the number concentration of small particles (less than $0.1\text{-}\mu$ radius) falls off with increasing altitude, consistent with a terrestrial origin, the concentration of the large (0.1- to 1-μ radius) particles actually shows a maximum at about 18 km, suggesting that these particles (predominantly sulfates with evidence of many nitrates) are formed in the stratosphere, as suggested by Junge (1963), who first identified this stratospheric phenomenon. The most generally accepted hypothesis for their origin is that they are formed by the oxidation of gaseous SO_2 and then grow, at least initially, as very hygroscopic sulfuric acid droplets (Cadle et al., 1970). Any cations would, of course, combine with the acid, but to what extent the number of droplets is reduced by such combination is not known.

The origin of the stratospheric SO_2 is of considerable interest.

We see very marked increases of the sulfate layer following large volcanic eruptions in the tropics, the most celebrated ones of recent history being the eruptions of Mount Tambora (1815), Krakatoa (1883), and Mount Agung (1963). Of course, there were no opportunities to sample the earlier stratospheric layers, but the characteristic height (about 18 km), persistence (several years), and twilight displays of colors leave little doubt that all major tropical eruptions have been alike in injecting large amounts of SO_2 (or possibly sulfates) into the lower stratosphere and that the particles that were subsequently formed spread worldwide in the surprisingly short time of about six months. The Agung particles had a measureable effect on the temperature of the stratosphere (see Section 1.4.7).

Typical concentrations of such stratospheric particles, measured several years after the Agung eruption and possibly representing a partial return to "normal," are as follows (Cadle et al., 1969; Cadle et al., 1970):

18 km in tropics	$0.25~\mu\,g/m^3$ ambient
18 km at mid-latitudes	$0.40~\mu\,g/m^3$ ambient
13 km at mid-latitudes	$0.15~\mu\,g/m^3$ ambient

(At 50 mb, or about 20 km, $1~\mu\,g/m^3$ ambient represents approximately 20 ppb by weight of sulfate particles.)

There is a first suspicion that man-made particles have begun to contaminate the stratosphere: the Cl/Br ratio for stratospheric particles is about 1/20 of its value for seawater, raising the possibility of a significant uptake of automobile exhaust particles that contain bromine (Cadle et al., 1970). The discovery of stable lead in the stratospheric population of particles would go a long way toward confirming this conjecture, and a systematic search for it is therefore in progress (Brookhaven National Laboratory, 1969).

Estimates of the present and future, natural and man-made production of particles is contained in reports of Work Groups 5, 6, and 7.

Lifetimes and Behavior of Particles

The removal of particles, both those that are soluble in water and those that are insoluble, is accomplished primarily by rain or snow. Very small particles are collected by small cloud droplets that are under the influence of Brownian motion, and subsequently the cloud droplets are rained out. Larger particles can be removed by the same two-stage process or by direct washout by falling raindrops.

The average lifetimes of particles in the lower atmosphere depends on the rainfall (or snowfall) regime in which they reside. Studies using various radioactive tracers have given lifetimes ranging from 6 days to 2 or more weeks in the lower troposphere. At mid-latitudes the shorter lifetime seems to be more accurate. In the upper troposphere the residence time is probably 2 to 4 weeks, while in the lower stratosphere the residence time varies from 6 months at high latitudes to about a year just above the tropical tropopause. The residence time continues to increase with altitude and is about 3 to 5 years in the upper stratosphere, and 5 to 10 years in the mesosphere, based on experience with specific radioactive nuclides that were injected as debris at high altitude by various nuclear tests (Junge, 1963; Telegadas and List, 1964; Bhandari, Lal, and Rama, 1966; Leipunskii et al., 1970; Martell, 1970.)

Because particles stay in the air for some time, and because there are a variety of different substances in the mix of aerosols, particularly over land, it is unusual for a particle to retain its physical identity. The accretion of foreign substances by a particle takes place by coagulation of particles with each other, by cycles of condensation and evaporation in water droplets, and by various gas reactions on a particle's surface.

One particularly important physical characteristic of particles is their ability to serve as nuclei for condensation of water vapor. Generally the larger particles (0.1 to 1 μ or larger) are the most effective condensation nuclei, but hygroscopic salts and acid droplets also serve in the same role. As the relative humidity increases in an updraft and approaches saturation, the more effective condensation nuclei start to grow first until from about 50 to several hundred droplets per cubic centimeters are formed, depending on the rate of cooling and the availability of nuclei. This is the mechanism by which warm clouds (most clouds, in fact) are formed. Whether they will precipitate depends very largely on the number of nuclei that were initially available in the updraft.

Another important physical characteristic of particles is

their ability to act as freezing nuclei that initiate the freezing of supercooled water droplets. A number of kinds of clay commonly serve as freezing nuclei, and there are other substances that are equally effective—including the celebrated silver iodide that is so popular with cloud seeders. Artificial sources of freezing nuclei that have been identified are steel foundries and some other plants that put out particles. The lead from automobile exhaust, when combined with any iodine that happens to be in the air, can be another large source in urban areas (Schaefer, 1966).

Trends

There is substantial evidence of an increasing trend in turbidity. The two longest records of turbidity, using standardized techniques at Washington, D.C., and Davos, Switzerland (McCormick and Ludwig, 1967), both show an upward trend in the past three or more decades. The Washington trend may be ascribed to the urban pollution nearby, but the Davos change, since it is recorded at a mountain observatory, may represent a more general increase in aerosol content.

Ludwig, Morgan, and McMullen (1969) showed that at 20 nonurban U.S. sites average ambient particulate concentrations increased by about 12 percent during the period 1962–1966, and suggested that this increase was the result of particles generated by man's activity. Schaefer (1969) found that at Yellowstone National Park and at Flagstaff, Arizona, Aitken nuclei concentrations increased by a factor of 10 within a five-year span during the mid-1960s.

Observations of direct solar radiation in North America and Europe (Budyko, 1969) show a radiation decrease of about 4 percent from the 1930s and early 40s to 1960. Budyko suggests that this decrease is due to atmospheric aerosols resulting from man's activity. The standardized network organized by the National Air Pollution Control Administration (NAPCA) has been in existence long enough to establish geographical patterns of total particulate loading, but not long enough to establish secular trends definitely.

Although very little is known about trends in turbidity far from sources of man-made particles, there are two sets of data of atmospheric electrical conductivity over the oceans that do give an indication. The conductivity of the lower atmosphere can be measured with good absolute accuracy from ships and is roughly inversely proportional to the particulate loading. Series of such measurements were made by the research vessels *Carnegie,* starting in 1907, and *Oceanographer* in 1967. Comparison of the two series (Cobb and Wells, 1970) showed no change in the South Pacific data, but the North Atlantic data indicated an increase in particle concentration by about a factor of 2, presumably due to pollution from the North American continent which persisted after it had been carried far out to sea.

Conclusions

1.   The stratosphere, because of the long residence times of contaminants and the photochemical processes there, is more sensitive to the injection of particles than is the troposphere. There is some evidence that man-made contaminants (for example, bromine) released at the ground are accumulating in the stratosphere in trace amounts. The major source of stratospheric particles at present is volcanic, consisting primarily of sulfates and nitrates, the former usually predominating. It is not known what the nonvolcanic fraction is, or the amounts of cations (ammonia, metallic ions) that are available to combine with the sulfates and nitrates. There is still some question about the magnitude of the increase in stratospheric particles following the last major eruption (Mount Agung in 1963), but it was probably more than ten times on a worldwide basis.

2.   The current quantitative yields of the principal sources of tropospheric particles that affect turbidity ($r \leq 2.5$ μ) have been estimated. The man-made contribution averaged over the globe amounts to about one part in five by weight, and the ratio of these number concentrations is about the same. The ratio is expected to rise. The overall trend in particles has been found:

a. To be increasing as shown by European and American solar radiation data (Budyko, 1969).

b. To be increasing over the North Atlantic, as shown by conductivity measurements (Cobb and Wells, 1970).

c. To be increasing at nonurban U.S. stations, as shown by high-volume filter samples (Spirtas and Levin, 1970).

d. To be increasing at Yellowstone National Park and at

Flagstaff, Arizona, as shown by Aitken nuclei measurements (Schaefer, 1969).

e. Not to be increasing significantly in data collected in the central Pacific (Cobb and Wells, 1970).

3.   In view of their relatively short residence times, the population of tropospheric particles can maintain large inhomogeneities. Problems of climate modification due to changes in this population will therefore tend to become manifest on a local level long before they do so on a global scale. Furthermore, the short residence time permits man to reverse the trend of growing atmospheric particle burdens within a few months if control measures are employed.

Recommendations

We recommend as priority efforts in order to identify particles and their trends (in addition to those recommendations concerning *effects* of turbidity contained in Section 1.4—also see report of Work Group 3):

1.   Monitoring at sites not biased by proximity to pollution sources of:

—Atmospheric turbidity; the minimum net of coverage being the same as that for CO_2. (For economic reasons they should as far as possible be the same set of stations.)

—Direct and global (all-sky) solar radiation, using standardized observations, with accuracy consistent with determining long-term trends. These should include measurements at several wavelengths when possible.

—Nature and concentration of particles (a) *in the stratosphere,* to support source identification, (b) *in the troposphere,* to support, in addition, an understanding of radiation and cloud physics effects.

—Trace gases that are the precursors of particles.

2.   Laboratory and field research on detection of specific man-made particles contaminating the stratosphere. Examples of such research are: a search for lead, the determination of the ammonium-ions/nitrates ratio among nitrogen compounds, full characterization of volcanic particles, determination of the sulfur-32/sulfur-34 ratio of sulfates, and so on.

1.2.3

Heat Released

The waste heat that is cast off during the processes of energy generation and consumption is as much a climatic contaminant as are the gases and particles that we put into the atmosphere. We know that a concentration of heat sources (for example, a city) affects local climate (Peterson, 1969). The hard question is, At what point will our waste heat become a global climatic factor? Our ability to answer this question is hampered by the necessarily crude estimates of future power generation amounts and locations and by our still-uncertain understanding of the dynamic process by which this heat would affect gobal climate.

It is possible to calculate the approximate magnitude of the heat released now and over the next 30 years. Greenfield (1970), using current energy utilization figures and the assumed rates of growth listed in Table 1.3, estimated that thermal waste power of the world will rise from 5.5×10^6 megawatts in 1970 to 9.6 $\times 10^6$ megawatts in 1980, and will reach 31.8×10^6 megawatts by 2000 (see Table 1.3). It should be noted that as a general matter all rates of growth are extrapolated from past history and are questionable when used to project much beyond five years. This is due to the fact that we are not yet capable of taking into account the effect of factors such as changing national goals and environmental constraints on power usage.

Lees (1970) has examined the 4,000 square mile area of the Los Angeles basin and has calculated that, at the present time, this area generates thermal power equivalent to more than 5 percent of the solar energy absorbed at the ground. He estimates that this will rise to about 18 percent in 2000 (see Figure 1.3).

A similar calculation was made (Greenfield, 1970) for a larger area, in the northeast section of the United States, where 40 percent of the national energy utilization occurs. In this area (roughly 350,000 square miles) the thermal waste is currently equal to approximately 1 percent of the absorbed solar energy and is projected to reach 5 percent by the year 2000 (see Figure 1.4).

Waste thermal power is available to the atmosphere as either sensible or latent heat. The latent heat is produced primarily in

Table 1.3
Thermal Waste Energy
(in units of 10^6 MW)

Geographic Location	Assumed Yearly Increase (percent)	1970	1980	2000
World	5.7	5.5	9.6	31.8
North America	4	2.2	3.4	7.5
(United States)[a]	(4)	(2.0)	(3)	(6.5)
(Canada)	(7)	(0.183)	(0.36)	(1.0)
Central America	6	0.12	0.2	0.68
South America	6	0.09	0.16	0.5
Western Europe	4	1.08	1.6	3.5
Western Asia	10	0.05	0.13	0.81
Far East	10	0.44	1.1	7.2
Oceania	8	0.069	0.145	0.64
(Australia)	(8)	(0.06)	(0.13)	(0.58)
Africa	6	0.1	0.18	0.57
East Europe	7	1.37	2.7	10.4
(Russia)	(8)	(0.98)	(2.0)	(9)

Source: United Nations, *World Energy*   [a] (   ) indicates subregion of main region.
Supplies.

the generation of electrical power and is dissipated by deposition in water bodies or by circulation through cooling towers. It should be noted that the latent heat will generally be available to the atmosphere at higher altitudes than the sensible heat and probably at some distance from the source. Based on the fractionation between waste heat from electrical power generation and from other sources, we may estimate from Figure 1.3 that, in 1970, 15 to 20 percent of the total waste heat will enter the atmosphere through the evaporation–latent heat route in the Los Angeles area, and that this will increase to approximately 30 percent by the year 2000. There is no apparent reason for not applying this estimate to other areas of the country as well.

1.2.4
Jet Aircraft Contributions to the High Atmosphere
There are two aspects of jet aircraft operations that deserve attention as possible sources of pollution, and consequent weather and

Figure 1.3 Thermal Power Generation in the Los Angeles Basin

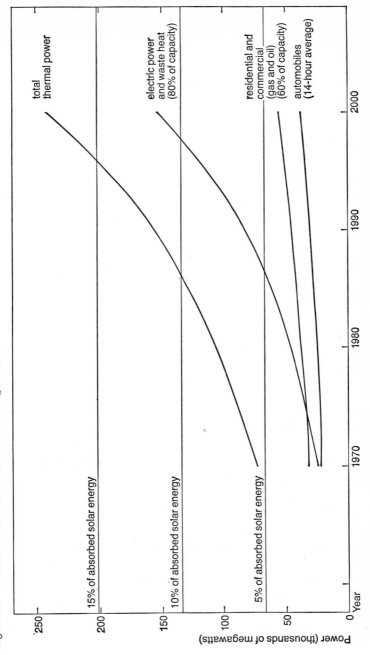

Source: Lees, 1970

Figure 1.4   Thermal Power Generation in a Climatically Significant Area in the United States

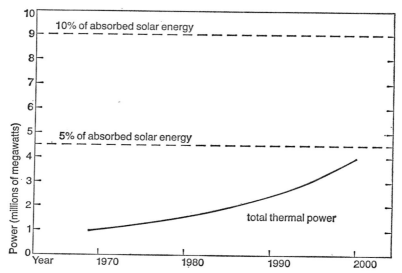

Source: Greenfield, 1970
Area: East, North Central States (Mich., Ill.; Ind., Wis.; Ohio); Middle Atlantic States (N.Y., N.J., Pa.) 351,028 mi²
Assumptions:
1. Electric power + waste heat = 30 percent of all power
2. Electrical capacity increases by factor of 10, 1970–2000
3. All other energy doubles, 1970–2000

climate modification: the condensation trails ("contrails") formed in the upper troposphere by subsonic commercial jets, and the contamination of the stratosphere by supersonic transports (SSTs) flying at about 20 km (65,000 feet).

Cirrus from Jet Contrails

The relative humidity, or proximity to saturation of water vapor, tends to increase with altitude in the troposphere so that the upper troposphere is most vulnerable to cloud formation from the addition of moisture. Also, the average residence time of a water vapor molecule is longer in the upper troposphere than in the lower troposphere; therefore, on the average, more water vapor can accumulate before being removed by precipitation. For these reasons, contrail formation, so common near the world's air routes, is more likely to occur when jets fly in the upper troposphere than when they fly in the lower troposphere.

There are very few, if any, statistics that permit us to determine whether the advent of commercial jet aircraft in about 1958 has altered the frequency of occurrence, or other properties, of cirrus clouds. Appleman of the U.S. Air Force Air Weather Service (1965) analyzed the trend of cirrus cloudiness on days with five-tenths or less low and/or middle cloudiness at 30 U.S. and non-U.S. Air Force bases, from about 1949 to 1965. He found a general increase in cirrus clouds during that period, which may not be statistically significant. Machta and Carpenter (1970) have analyzed the trends of cirrus cloudiness at Salt Lake City and Denver between 1949 and 1969, again for days with no or few obscuring lower clouds. The results, still very tentative, indicate that cirrus cloudiness has increased. These results are statistically significant. They noted that the cirrus cloudiness increase paralleled the increase in jet aircraft activity (as measured by the amount of fuel used by domestic airlines). However, slight decreases in middle cloudiness at Denver suggest that observers may have relabeled some of the middle clouds as upper clouds after jet pilots reported their true heights. It should also be noted that all stations whose records have been examined lie near or under commercial air routes.

Estimates of future subsonic jet operations, while uncertain, indicate perhaps a three- to sixfold increase in the years 1985–1990. These estimates would probably have to be increased if the SST, which would cruise in the stratosphere, instead of the troposphere, does not come into operation.

Recommendations

We recommend that the magnitude and distribution of increased cirrus cloudiness from subsonic jet operations in the upper troposphere be determined. A study of the phenomenon should be conducted by examining cloud observations at many weather stations, both near and remote from air routes.

SST Contamination of the Stratosphere

Contamination from the future operation of SST (supersonic transport) aircraft requires special attention because of the portion of the atmosphere, the stratosphere, in which the contamination will be injected. The stratosphere differs from the troposphere, for our purposes, in two important ways: (1) contaminants will remain airborne for longer periods there (1–3 years compared to several weeks), and (2) the stratosphere may be sensitive to contaminants because concentrations of parts per million or less may participate in its photochemistry.

Of all the SST combustion products, carbon dioxide is the most abundant. The average life of a carbon dioxide molecule in the air before passing into the sea or the biosphere is about 5 years (estimates vary from about 3 to 30 years). This is appreciably longer than the 2 years that we shall use for the stratospheric residence time of an inert gas inserted at about 20 km (60,000–70,000 feet) by the SST. There would therefore be time for mixing of CO_2 to occur between stratosphere and troposphere. Further, during the 5 years, 1985–1990, atmospheric carbon dioxide would increase by about 0.5 ppm (by volume) from SSTs, and by about 10 ppm from other fossil fuel combustion at the ground (see Table 1.2). For both reasons, the contamination of the stratosphere by SST carbon dioxide does not appear to be as important as the contamination of the whole atmosphere by other, much larger sources of carbon dioxide.

The second most abundant product of the combustion of jet fuel is water vapor. In contrast to the wet troposphere, the stratosphere is exceedingly dry, containing only about 3 ppm (by mass) of water vapor. It is therefore possible to alter stratospheric humidity by SST activities. Indeed the proposed SST flights for 1985–1990 (Federal Aviation Administration, 1967) would increase the global average stratospheric concentration of water vapor by 0.2 ppm (from 3.0 to 3.2 ppm). However, this increase will be nonuniform, ranging from high values directly behind the aircraft to small increases in the south polar regions. Pending more rigorous calculations of the likely distribution, it will be assumed that significant volumes of the stratosphere, in terms of climate change, may be as much as 10 times (that is, 2 ppm) higher than before SST operations.

Carbon monoxide and nitrogen dioxide (a likely daughter product of nitric oxide) both absorb infrared terrestrial radiation.

However, their small concentrations (parts per hundred million) and small absorption rates make it highly unlikely that either can directly compete with ozone, carbon dioxide, or water vapor in altering the stratospheric radiation budget significantly.

Both carbon monoxide and nitrogen in its various oxide forms can also play a role in stratospheric photochemistry, but despite greater uncertainties in the reaction rates of CO and NO_x than for water vapor, these contaminants would be much less significant than the added water vapor and may be neglected.

The currently accepted theory accounting for the predominance of sulfate in the stratospheric particulate layer (the Junge sulfate layer, at approximately SST cruising altitudes) argues that SO_2 from the lower atmosphere enters the equatorial stratosphere as a gas, where it is converted photochemically to particulate sulfates. Thus, direct introduction of SO_2 into the stratosphere would increase the amount of sulfate produced. Furthermore, reactions that occur in the lower atmosphere between NO and hydrocarbons can also take place in the stratosphere, converting hydrocarbons to particles in the presence of sunlight (the "smog" reaction). Unfortunately, the efficiency of these reactions in the stratospheric environment is uncertain. For the present purposes, we shall assume that *all* SO_2 is converted to sulfate particles and that the entire mass of emitted hydrocarbons is converted to organic particles (see Table 1.6 in the Appendix to this section). Some of the NO_x could be converted to nitrate particles but we believe that this conversion will be very small due to the unavailability of suitable positive ions and the fact that nitric acid is not hygroscopic.

The natural background of particles was measured in the early 1960s by Junge (1963) and again in the late 1960s by Cadle et al. (1970), who derived concentrations about thirtyfold higher. This increase is probably due to added particles injected into the stratosphere by the eruption of Mount Agung in 1963 and to subsequent tropical eruptions, but the large factor may also be partly due to inefficient collection systems used by Junge et al. (that is, the earlier numbers may be too low). Thus, it is likely that the true, nonvolcanic sulfate background lies between the two values shown in Table 1.6.

Because one cannot verify the stratospheric radiation balance using gases alone in the calculation, it is believed that the Junge sulfate particle layer may play a role in the heat balance of the lower stratosphere (Junge, 1963). Newell (1970a), for instance, has shown that shortly after the injection of debris by the volcanic eruption in March, 1963 of Mount Agung in Bali, the temperature of the lower equatorial stratosphere rose about 6° or 7°C, presumably because of the absorption of solar radiation by the injected particles at that altitude. The position will be taken in this discussion that weather and climate effects are possible when the increase in concentration due to added particles is of the same order as the natural post-Agung concentration of particles in the Junge sulfate layer.

Table 1.6 of the Appendix to this section shows that the global average values for sulfates, hydrocarbons, and soot are about equal to or greater than the Junge pre-Agung values; and that the peak north temperate latitude concentrations of these particles approach or (in total) exceed even the post-Agung values.

Although the foregoing calculation is conservative in the sense that we have tried to avoid any overestimate of the formation of sulfate and hydrocarbon particles from SSTs, it is possible that the artificial particulate concentration might be larger than our estimate if: (1) the NO_x, which is produced in rather appreciable quantity, were to form particles as well as the SO_2, and thus increase particle production by a large factor; (2) the GE-4 turns out to approach present-day large-engine emission rates of soot and hydrocarbons which could raise the particles originating as both soot and hydrocarbons by over a factor of 5. Furthermore, it should be pointed out that the particle formation rates observed during static ground-level tests of jet engines, or derived from theoretical calculations, may underestimate the particle formation rate which will occur when the exhaust products condense in the very cold ($< -65°C$) stratospheric environment.

On the other hand, if it were economically feasible to produce and use very low sulfur content fuel, namely, 0.01 percent S, instead of continuing with that currently available (0.05 percent S by mass), then the sulfate contribution would be cut to one-fifth of that shown. However, the total particulate addition would still be at least half of that estimated in Table 1.6, since there would be no reduction in the amount of soot and converted hydrocarbons. The total particulate addition would also still be comparable to the apparent, pre-Agung, global particle background.

Appendix: Statistics on SST Engine and Calculations of Stratospheric Contamination

The emissions from a single GE-4 engine to be used on the B2707-300, flying at 65,000 feet (about 20 km) in a cruise mode, at Mach 2.7, on a day with standard weather conditions, have been calculated (Table 1.4), assuming chemical equilibrium, by General Electric Company engineers (Hession, 1970; Thompson, 1970). In the future, the calculations will be compared with actual engine tests, but it is estimated by SST officials (Thompson, 1970) that the fraction of products per unit amount of consumed fuel may be accurate within 10 percent (except as indicated later). It should be noted that the engine will be equipped with an afterburner; however, it will be in only partial operation in the stratosphere, as opposed to full operation during takeoff; that is, the temperature in the afterburner will be 2,800°R (degrees Réaumur) while cruising in the stratosphere, but will be 3,500°R during takeoff.

An estimate suggested by the Federal Aviation Administration (FAA) calls for 500 SST aircraft to be flying during the period 1985–1990. Each of these jets would fly in the stratosphere for almost 7 hours a day (2,500 hours per year). Of the 500 jets, 334 would be U.S. and be equipped with four engines, and 166 would be of non-U.S. fabrication and would have the equivalent of two engines.

Tables 1.5 and 1.6 contain our estimates of equilibrium or steady-state concentrations, after several years of operation, assuming a 2-year mean residence time for all products.

Conclusions

The following conclusions are based on the assumed operation of 500 commercial SSTs in the period 1985–1990, and on their emissions as provided by the federal SST office (except as noted).
1. Because of the long atmospheric residence time of CO_2 and

Table 1.4
Statistics of Emissions from One GE-4 Engine, Cruise Mode

Constituent	Pounds per Hour
Ingested air and consumed fuel	
air	1,380,000
fuel	33,000
Unused air	
N_2	1,039,000
O_2	208,000
Ar	19,300
Combustion products	
CO_2	103,500
H_2O	41,400
CO	1,400
NO[a]	1,400
SO_2[b]	33
Soot (Particles)	5
Unused fuel	
Hydrocarbons[c]	16.5

[a] The General Electric Company advises (Thompson, 1970) that the true NO output is likely to be no more than one-half to one-third of the calculated value. A few past comparisons suggest that measured values will be values that are 10 to 15 percent of the calculated numbers (Thompson, 1970).

[b] Sulfur content of SST fuel will be specified as no more than 0.3 percent sulfur by weight. The 33 pounds per hour given in the table corresponds to 0.05 percent sulfur by weight, which is the average sulfur content of currently available jet fuel, as de- termined by the Department of the Interior. An appreciable amount of jet fuel with less than 0.016 percent S is available now, according to a Boeing Company survey (Swihart, 1970). A leading producer has told us that it is technologically feasible to produce jet fuel with only 0.01 percent S, but that present production facilities are not adequate to supply a large market.

[c] The General Electric Company expects the true hydrocarbon emission to be vanishingly small, due to the high temperature of the afterburner (Thompson, 1970).

the relatively small CO_2 contribution from SSTs compared with that from other sources of fossil fuel combustion, there will be no special CO_2 problem due to SST operations.

2.   Stratospheric water vapor will increase, on a global average,

Table 1.5
Gaseous Concentrations

	10^{11} Grams/Yr[a]	World Average Concentration[b]	Possible Peak Concentration[c]
CO_2	1,960.	Not relevant	
H_2O	783.	0.20 ppm (0.3)	2.0 ppm (3.0)
CO	26.3	6.8 ppb (7.3)	0.068 ppm (.073)
NO	26.3	6.8 ppb (7.3)	0.068 ppm (.073)
SO_2	0.63	0.16 ppm (0.20)	1.6 ppb (2.0)
Hydrocarbon	0.31	0.081 ppb (0.087)	0.81 (0.87)

[a] This column is computed using data from Table 1.4 and assumptions for full-scale SST operations for 1985–1990 given in the text.

[b] This column is computed using the data from the first column and the assumption of a two-year mean residence time for all products in this portion of the stratosphere. The mass of the atmosphere is taken as 5.14×10^{21} grams, with the stratosphere containing 15 percent of that, or 0.77×10^{21} grams. The numbers in parentheses represent rounded values obtained by increasing the SST-produced concentration to accommodate subjective estimates of military injections of contaminants into the stratosphere. All concentrations are by mass.

[c] The last column is obtained by multiplying the world average by 10 and represents a tentative upper limit for certain temperature change considerations that would apply to a region of high SST activity.

by 0.2 ppm by mass (from 3.0 to 3.2 ppm). Since there will be more water vapor added to the north temperate latitudes, parts of this region may perhaps have a contribution to standing concentration as much as tenfold higher than the increase of the global average (that is, grow from 3 to 5 ppm of water vapor).

3.   Concentrations (by mass) of CO, NO, SO_2, HC, and soot will range from fractions of a ppb for soot to 68 ppb for CO and NO from 500 SSTs in the north temperate latitude, peak of SST activity. Emissions from present large engines are appreciably different from those predicted for the SST GE-4 engines. The current NO emission rates per gallon of fuel for certain jet engines are as much as twentyfold lower, but the HC can be fourfold greater and the particles more than tenfold greater. Fuel with a 0.3 percent sulfur content (the maximum allowable limit) would increase the emission of SO_2 by more than sixfold over our SST estimates, and use of 0.01 percent sulfur fuel (technically feasible)

Table 1.6
Particle Concentrations
(10^{-4} ppm by mass)

| | Formed from SST Exhaust | | Measured at SST Altitude | |
	Global Average[a]	Peak N. Hem.[b]	Pre-Agung,[c]	Post-Agung[d]
SO_4^-	2.4 (4.8)[e]	24 (48)	1.2	36
Hydrocarbon	0.81 (1.6)	8.1 (16)	——	——
Soot	0.25 (0.5)	2.5 (5)	——	——
Total	3.46 (6.9)	34.6 (69)	1.2	36

[a] The SO_4^- figure was calculated using data in Table 1.5 and assuming all SO_2 is converted to SO_4^-. (Note that SO_4^- is heavier than SO_2.) The hydrocarbon figure is taken from Table 1.5 and assumes total conversion of gas to particles. The soot figure is computed using data from Table 1.4 and assumptions for full-scale SST operations for 1985–1990 given in the text.

[b] A factor of 10 is applied to the first column in order to estimate the peak concentration within the nonuniform distribution over the globe and would apply to the regions, mostly in the Northern Hemisphere, of large SST activity.

[c] Source: Junge, 1963. No data exist for carbon or hydrocarbon concentrations in the stratosphere.

[d] Source: Cadle et al., 1970

[e] Numbers in ( ) are twice the calculated value, to account for particle settling. It is believed that sulfate particles, because of their slow settling speeds, remain concentrated in the sulfate layer rather than disperse throughout the vertical extent of the atmosphere. The natural sulfate layer happens to be at about the same altitude as the SST injections. The net effect is probably an increase by about a factor of 2 over the concentration calculated for a gas, which would diffuse upward as well as downward. This factor of 2 has been applied in Table 1.6 (the values in parentheses) to the calculated concentrations of all three particles—sulfates, hydrocarbons (HC), and soot—for the same reason.

would reduce the SO_2 emission to one-fifth of our estimates, since the calculations here are based on 0.05 percent sulfur content. Further, there are doubts concerning the applicability of particle emission information derived from both theoretical calculations and static field tests to the real atmosphere and operating conditions. It is felt that realistic operation may produce larger numbers of particles. Finally, we have no information on fuel additives.

1.2.5

Oxygen: Nonproblem

Atmospheric oxygen constitutes about 21 percent of the atmosphere by volume. There is virtual equilibrium between photosynthetic production of it and its utilization by animals and bacteria. The mean residence time may be of the order of 10,000 years (Johnson, 1969). From time to time there have been predictions of a disruption of this balance; most recently it has been speculated that there may be a reduction in photosynthetic organisms in the ocean by herbicides and pesticides. To obtain a current base-line point in order to detect any (unlikely) changes in atmospheric oxygen, Machta and Hughes (1970) measured oxygen concentrations in a large number of clean air locations. The mean value was found to be 20.946 percent by volume with virtually no discernible geographic variation. The accuracy of the comparison standard was ± 0.006 percent by volume (one standard deviation).

Comparison with measurements taken since 1910 suggests no detectable change in concentration. The best earlier measurements also indicate 20.946 percent by volume. All others that are reliable lie within the uncertainty of the measurements. Past data show only a negligible lowering of oxygen concentration in the most polluted open areas.

Finally, calculations of the consumption of oxygen that would be caused by the burning of all known recoverable fossil fuels indicate that oxygen concentration would be reduced to 20.800 percent by volume. Such a reduction would have a negligible effect on human or animal respiration.

Conclusion

There has been no detectable change in atmospheric oxygen in this century. Future changes, if any, will occur very slowly.

Recommendation

We recommend that monitoring be conducted only at very infrequent intervals, such as every 10 years, in light of the theoretical and observed stability of atmospheric oxygen.

1.2.6

Radioactive Nuclides

Estimates of the production of various radioactive waste products by the expanding use of nuclear power are made in the report of Work Group 7. There is no question that two radioactive products will find their way into the air and water system of the earth: tritium (as tritiated water), and krypton-85 (a chemically inert noble gas). Both have half-lives of about 12 years, so they will have a chance to diffuse throughout the globe regardless of the location of their sources.

The complex question of how much of these two substances we can tolerate in our environment is related entirely to their effects on living things, including man. We shall not deal with that here.

As far as possible effects of these radioactive substances on the physical processes in the atmosphere are concerned, we can identify none at the concentrations which have been projected to the year 2000. The atmosphere in this case will be far more resistant to these nuclides than the biosphere.

1.3
Changes of the Earth's Surface

The most important properties of the earth's surface that have a bearing on climate and are likely to be affected by human activity are the following:

1.   Reflectivity. When the reflectivity (albedo) of the ground for shortwave radiation is modified, the amount of sunlight returned to space will be changed along with the energy available to heat the ground and the atmosphere just above it. Thus, decreasing the surface albedo usually raises the ground temperature, and, therefore, the air temperature immediately above the ground as well. The result of that effect is that the overall lapse rate, that is, rate of decrease of atmospheric temperature with altitude (not limited to the region immediately above the ground) is increased. The atmosphere becomes more unstable, and thus experiences increased vertical mixing.

2.   Heat Capacity and Conductivity. The smaller the heat conductivity and capacity of the surface, the larger will be the surface temperature during the day and the greater will be the lapse rate and the vertical dispersion of contaminants and, consequently, atmospheric properties.

3.   Availability of Water. Increased availability of water on the surface will increase the portion of the sun's energy that is used in evaporation. Therefore, less of the sun's energy will be used in surface heating, and the surface temperature will be cooler. In addition, the water vapor produced by this increased evaporation will be available in the atmosphere to form clouds and to cause precipitation downwind from the water source.

4.   *Availability of Dust.* Human activity may change the mechanical properties of the surface so that more or less dust can be carried from it into the atmosphere. The effect of such dust is of potentially great importance and is dealt with in Section 1.4.6.

5.   *Aerodynamic Roughness.* The rougher the ground, the weaker will be the wind above it and the stronger will be the turbulence of the air (irregular fluctuations of the wind, both in the vertical and in the horizontal). Therefore, for a given lapse rate, vertical mixing will be stronger over rough terrain than over smooth terrain—for example, vertical mixing will be stronger over a city than over a water body.

6.   *Emissivity in the Infrared Band.* The emissivity of the surface in the infrared band is always nearly 100 percent, but the small change in emissivity produced by a change in the character of the surface may have some significance.

7.   *Heat Released to the Ground.* Man's activities produce large amounts of heat that is released at or near the earth's surface. This matter is dealt with in Sections 1.2.3 and 1.4.6.

1.4
Theory of Climate and Possibilities of Climatic Change

While the main thrust of the Study has been to determine where man may be influencing and changing his environment adversely on a global scale, we must not forget that the atmospheric environment has changed many times during the history of the earth. Man might now be forcing his way with or against the natural tide of climatic change. We do not know whether such a powerful tide can be stemmed, but man could quite conceivably abet it. In any case, we must take naturally occurring changes into account when we think of the future decades and centuries, since they have had a major impact on mankind in the past and promise to do so again—sometime in the future.

In the following sections, some of the lessons of the past are reviewed so that we may proceed, with a better perspective, to a discussion of man's present influence. It will be also clear that we need to know much more about the natural forces that are at work, at work more inexorably, perhaps, than we realize.

1.4.1
Mathematical Models of the Atmosphere

On both practical and conceptual grounds, the use of mathematical computer models of the atmosphere is indispensable in achieving a satisfactory understanding both of climatic changes, as we observe them, and of the potential role that man may play in causing them. The conceptual reason is that, without knowledge of the alternative behaviors with which the atmosphere could have consistently responded to the changing conditions of the planet, we shall be unable to ascribe observed effects unambiguously to possible causes. From a practical viewpoint, the dynamic behavior of the earth's fluid envelope presents us with so many interdependent phenomena and processes, governed by so many feedback cycles (some stabilizing and others destabilizing) that the only way that we now conceive of exploring the tangle of relations which describes all this is the numerical solution of specific examples.

Corresponding to this pair of focal reasons for the mathematical simulation of global atmospheric behavior are two characteristic uses of such computer models. First, they serve, one might say, as the apparatus for conducting laboratory-type experiments on the atmosphere-ocean system which are impossible to conduct on the actual system. Used in this mode, models contribute to our understanding of climate dynamics. Second, models are run in such a way as to produce results of possible operational usefulness, such as longer-term forecasts of global atmospheric conditions.

Global models of the earth's fluid envelope, to be of any value, are necessarily elaborate and therefore costly to design and operate. Nevertheless, our dependence on them for making progress in this field is such that the effort of developing them cannot be avoided.

Model Design

Mathematical models of the atmosphere are derived from consideration of the atmosphere's behavior as an aerothermodynamical process. A minimum of seven quantities are required to describe its state at any point: pressure, temperature, density, the three components of the velocity vector, and moisture (usually the fraction, by weight, of water vapor). These quantities and their time rates of change are related by four basic laws of physics (the "primitive equations"): Newton's second law (stated in two relations—for the horizontal motion and the hydrostatic approximation in the vertical); continuity of mass (two relations—one for total mass and one for water mass, to allow for the latter's change of phase); the first law of thermodynamics; and the equations of state (gas law). To these equations must be added the relations that define boundary constraints and external influences, especially the incidence of solar radiation, the flux of outgoing infrared radiation, and turbulent transport of heat, water vapor, and momentum at the lower boundary in order to complete this, in every respect primitively minimal, model of the atmosphere.

Although current numerical models of the general circulation are satisfactory for quite a few purposes, they run almost immediately into a number of difficulties because they fail to specify the behavior of the atmosphere completely. For example, they make allowance for the condensation of water vapor at specified states of that gas, but they do not determine whether the condensed water remains suspended in the form of clouds or falls out as rain. Another example is found in the first law of thermodynamics, which contains a term measuring the heat added to the air. This term depends on the latent heat budget and on radiational effects. Even if the former is specified as a function of the state of the air, the latter can cause difficulty by its dependence on the composition of the air. Eliminating this difficulty, by assuming the compositions of the atmosphere to be constant, automatically builds a major element of what might be called "now-climate" into the models. A third deficiency is that the variable effect of cloudiness on the radiation budget is not treated explicitly in current models.

A similar set of variables and equations can be used to describe the ocean, using salinity instead of moisture as the main variable for characterizing composition. Comparable degrees of incompleteness affect present ocean models, which similarly limit the usefulness of the models in simulating climatic changes. The atmospheric and ocean models can in principle be coupled, and in present designs are coupled, by making them respond to each other across their interface (that is, by requiring continuity of heat, momentum, and mass flow at each point).

Model Operation

A model is run by numerically integrating its equations, starting with some initial atmospheric regime. For this purpose the size of the time steps must be properly related to the spacings between points in the grid of discrete points that replaces the continuous atmosphere. If the steps are not properly related, the propagating computational errors, and not the conditions of the problem, will determine the computed results. This constraint takes the form of a practical ceiling on the time step, which, in the case of our atmosphere, is about 5 minutes for a horizontal grid spacing of 300 km, and decreases proportionally with the latter. Clearly, the integration of even a simplified version of such a model over any length of time becomes a formidable undertaking. Advanced models require two or three hours (or more) for each model day, using the largest computers available.

It is becoming increasingly clear that the atmosphere is capable of motions of practically all scales, varying from millimeters to thousands of kilometers and from a few seconds to millennia. The operation of a single model that would simulate all of these processes is therefore completely beyond us. We must be satisfied with models which are reasonably faithful over the range of scales that are of interest for a particular purpose. This poses the problem of how to simulate what might be called the leakage of energy and momentum to and from the scale ranges that lie beyond our simulation capability. Up to a point, various forms of ad hoc lumping and aggregation have provided satisfactory parameterizations of the interaction of model quantities with processes that lie below the scale of the model. However, there is a vexing dearth of such techniques at larger scales. For example, the very important effects of cumulus convection, which account for a major part of the vertical exchange of heat, moisture, and momentum in the tropics, are still not satisfactorily parameterized in current models.

Characteristics of Numerical General Circulation Models

Up to now we have been describing the kind of general circulation models that have been developed to simulate the atmospheric motion in some detail and that lend themselves to making forecasts for a few days when real initial conditions are applied.

In order to create a model that can simulate the climate and its long-term changes, it becomes essential to introduce some factors that do not have to be included in the shorter-term simulations, the ocean circulation being the most important. An attempt to run a combined ocean-atmosphere model has been made (Manabe and Bryan, 1969) but it had to be a highly simplified one so that it could run long enough for the ocean to approach some sort of equilibrium with the atmosphere—several centuries at least. The model did simply account for the variation of snow cover and sea ice but not variable cloud cover and its effects on the radiation budget. Variable cloud cover has not been satisfactorily handled in any numerical model so far.

The problem of estimating climate modifications that result from the rather puny activities of man is even more demanding. It requires models that are capable of simulating the natural climate and its variability in sufficient detail, and that must consider the effects of changing aerosols as well as carbon dioxide and energy inputs at the surface. Progress toward that end has been made, but the objective is far from attainment. It is not believed that models now operating can simulate the difference between the global climate of 100 years ago and today, because they do not take into account the long-term factors that we have mentioned.

Most of the inadvertent climate changes which man might cause represent variations that are comparable to the noise level of the natural variability of the atmosphere, as well as being within the uncertainty range of current modeling techniques. The real atmosphere is apparently capable of anomalies that last longer than a year—we hardly know how long—and some models appear to show the same characteristic. To obtain discriminating statistics will require long runs with more complete models, a rather costly proposition.

At a more fundamental level, however, the goal is to free progressively the models that we operate from their dependence on empirical relations derived from the climate as we now happen to find it. At the distant, probably unattainable, end is the replacement of those relations by an application of the basic laws of physics, resulting in models that would simulate with equal facility the climates of Earth, Venus, or Jupiter. It is more likely that we shall have to be satisfied with replacing empirical relations by increasingly general climatological laws, toward which we shall gradually grope our way as we amalgamate not only observations on the present climate but also the experience that we acquire from working with ingenious models, explicitly devised for this purpose.

Recommendations

We recommend that current models be improved by, first, including more realistic simulations of clouds and air-sea interaction and, second, that attempts be made to include particles in the radiation balance calculation when their optical properties become better known.

Such models should be run on many time scales, but at least for periods of several simulated years. With ocean circulation included in the atmosphere-ocean system, it will take several simulated centuries even to approach a quasi-steady state. Possibilities should be investigated for simplifying existing models to provide a better understanding of climatic changes. Simultaneously, a search should be made for alternative types of models which are more suitable for handling problems of climatic change.

1.4.2

Effects of Changing CO_2

If we assume that the earth-atmosphere system is in radiative equilibrium with the sun (that is, that it emits as much radiation as it receives), we can define an equivalent temperature ("brightness") for the planet as

$$T_e = \left[\frac{(1 - A)S}{4\sigma} \right]^{1/4}$$

where A is the planetary albedo, S is the solar constant and σ is the Stefan-Boltzmann constant. The factor 4 appears because the solar constant is defined for a surface perpendicular to the sun's rays, but the earth-atmosphere system radiates energy outward in all directions. For an average planetary albedo of approximately 33 percent (London and Sasamori, 1970), the radiative-equilibrium temperature of the earth-atmosphere system is 253°K (−20°C), a value that is very close to recent estimates from satellite observations (Raschke and Bandeen, 1970).

The average surface temperature T_s is higher than the earth-atmosphere radiative equilibrium temperature, as just defined, because the atmosphere is semitransparent for solar radiation but is fairly opaque for terrestrial (infrared) radiation. For instance, in the clear atmosphere (containing an average amount of water vapor, carbon dioxide, ozone, and dust), about 65 percent of the incoming solar radiation reaches and is absorbed at the earth's surface. For the same atmospheric conditions, however, only about 10 percent of the total radiation leaving the surface is directly transmitted back to space (that is, does not get absorbed by the atmosphere). For an average cloudy atmosphere these numbers are 45 percent and 5 percent, respectively (London and Sasamori, 1970). The remaining portion of the surface radiation is absorbed by trace atmospheric gases (primarily water vapor and carbon dioxide) and by clouds and is then reemitted downward to the ground and upward to space. The returned (downward) infrared radiation helps to keep the surface temperature relatively high.

The mean surface temperature is about 286°K (13°C). The ratio T_s/T_e is sometimes referred to as the "greenhouse" coefficient. For the earth this is $(286/253) = 1.13$. For Venus, the "greenhouse" coefficient is approximately $(700/265) = 2.60$.

As is well known, water vapor, carbon dioxide, ozone, and clouds are the principal active radiating substances in the earth's atmosphere. It is important, therefore, for present purposes, to comment on the importance of carbon dioxide, as compared with these other substances, in affecting the atmospheric radiation budget. Calculations show that in the troposphere both the radiative heating (solar radiation absorption) and cooling (infrared flux divergence) by water vapor is about ten times larger than that for carbon dioxide. In the vicinity of the tropical tropopause, infrared flux due to CO_2 actually leads to heating. This radiative convergence results because the carbon dioxide in the cold air at the tropopause "sees" warm air from above and below.

In the stratosphere, radiative heating due to the absorption of solar radiation by ozone is dominant. The radiative cooling that does occur is carried on primarily by CO_2. Water vapor plays a minor role above about 20 km.

In general, radiative processes in the atmosphere act to cool the atmosphere. Since, in the troposphere, the *net* radiative cooling is larger at higher latitudes (where the temperature is low) than in equatorial latitudes (where the temperature is high), energy is available for atmospheric motions in the troposphere (see, for instance, London and Sasamori, 1970). In addition, net radiative processes in the troposphere lead to maximum cooling rates at levels of 5 to 10 km. A vertical column of air tends toward hydrostatic instability when the air gets colder aloft relative to the air below. Thus, radiation acts to produce less hydrostatic stability in the lower troposphere, and greater stability above. The thermal radiative component due to carbon dioxide slightly reduces this generation of available horizontal and vertical potential energy (Newell and Dopplick, 1970).

Since carbon dioxide represents one of the components affecting the earth's radiation budget, changes in the carbon dioxide concentration (natural or man-made) will produce some changes in the radiation budget in general and in the surface temperature in particular.

It was first suggested by Tyndall in 1863 that the blanketing effect of increased carbon dioxide would cause climatic changes through variation of the surface temperature. Increased carbon dioxide, because of its strong absorption (and therefore emission) of infrared radiation at 12 to 18 μ, would reradiate energy downward to the earth's surface and further inhibit the radiative cooling at the ground. Water vapor, ozone, and clouds have qualitatively similar effects.

At the end of the nineteenth century, Arrhenius computed that a surface temperature rise of 9°C would result from a threefold increase of atmospheric carbon dioxide (Arrhenius, 1896). Renewed interest in this problem was sparked by the computations of Plass (1956), starting about 15 years ago, who used a fairly realistic model of the carbon dioxide 15-μ absorption band and high-speed computers to derive changes in surface temperature that would result from a change in carbon dioxide concentration from 300 ppm to 600 ppm or to 150 ppm. His calculations, assuming only the 15-μ CO_2 band, no overlap with water vapor absorption bands, and no surface evaporation or convention, gave a value for ΔT_s of $+3.6°C$ or $-3.8°C$, depending on whether the CO_2 concentration was doubled or halved, respectively.

Subsequent calculations have included clouds, the overlap of water vapor and carbon dioxide absorption bands, and the impact of increased water vapor (resulting from an increased surface temperature) as a feedback mechanism (Möller, 1963). Other computations included, in addition to the preceding, the effect of ozone and of a "convective adjustment" so that the lapse rate in the troposphere would not become unstable (Manabe and Wetherald, 1967). The influence of the absorption of near infrared solar radiation by carbon dioxide has been studied by Gebhart (1967).

The various results indicate that, if the surface heating due to CO_2 is accompanied by evaporation (to maintain a constant relative humidity), the "greenhouse effect" is amplified slightly. But if average cloudiness is included in the radiative model and/ or the overlap of water vapor and carbon dioxide infrared bands is accounted for, the surface radiative equilibrium heating due to an increase in atmospheric carbon dioxide concentration is damped to about one-half or less of what it otherwise would be. The average increase of surface temperature predicted by radiative equilibrium studies, which include a "convective adjustment" but no other dynamic or thermodynamic effects, is approximately 2°C for a doubling of the carbon dioxide concentration (Manabe and Wetherald, 1967).

In addition to the results of radiative equilibrium studies of the influence of carbon dioxide on the surface temperature, some computations have been made of the changes in the vertical radiative equilibrium temperature distribution and of the heating and cooling rates that would result from changes in the concentration of carbon dioxide (see, for instance, Gebhart, 1967; Newell and Dopplick, 1970). Figure 1.5, taken from Dopplick (1970), shows the contributions to the radiative heating and cooling produced by the principal trace gases involved: H_2O, CO_2, and O_3.

The results of these studies show that, if radiative processes

Figure 1.5   Radiative Heating Rates of Three Atmospheric Constituents

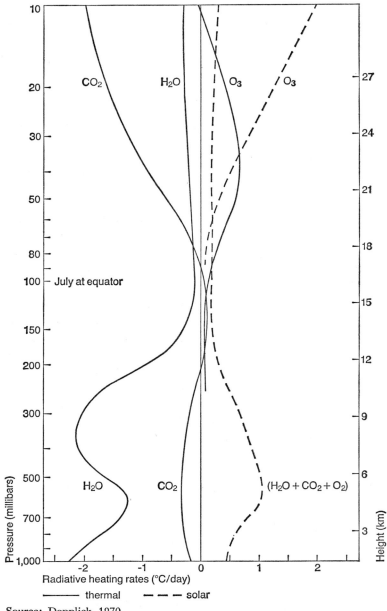

Source: Dopplick, 1970

alone are considered, the lower and middle troposphere would warm slightly and the stratosphere would cool if the carbon dioxide content were doubled. For example, stratospheric radiative equilibrium temperature would decrease by 2°C at 25 km and by about 12°C at 40 km. Radiative cooling at these levels would increase from about 1.0° to 1.1°C/day at 25 km and from 4.0° to 5.0°C/day at 40 km for an increase of carbon dioxide from 300 ppm to 600 ppm. The increased cooling (or lower radiative equilibrium temperature), if maintained when atmospheric motions are considered, would decrease the frost point required for water vapor saturation in the stratosphere and, conceivably, result in an increase in the occurrence and distribution of stratospheric clouds. It is probable, however, based on the computations of Manabe and Wetherald (1967), that these clouds would exert very minor influences (if any) on the temperature distribution in the troposphere. Both low-level warming and higher-level cooling would be reduced considerably at high latitudes during winter, as compared to summer.

The vertical and horizontal variation of radiative temperature changes due to carbon dioxide variations can affect the creation of atmospheric available potential energy, since radiative heating will occur in regions of the troposphere where it is relatively warm (lower troposphere at lower latitudes).

Thus, the radiative temperature changes should have an effect on the dynamics of atmospheric motions. However, it is important to note that the motions of the atmosphere interact in turn with the atmospheric temperature field generally in such a way as to reduce the temperature differences that exist.

In this discussion, and in all computations that have been made so far, three primary physical processes have been left out of the thermodynamic-hydrodynamic system that produces the aspect of climatic change being considered here. The three processes are the role of evaporation and condensation as energy transport mechanisms, the role of cloudiness (and the variation of cloudiness) in regulating the field of radiative flux in the earth-atmosphere system, and the role of the oceans in providing a heat reservoir to damp short-period temperature oscillations at the earth's surface. Although much physical insight can be (and has been) gained by the study of so-called radiative equilibrium models of the "greenhouse atmosphere," no definitive statement can be made regarding the influence of carbon dioxide variations on the temperature distribution of the atmosphere, or their effect on climate, until the radiative and dynamic effects are combined.

Conclusions

Radiative equilibrium computations, including a convective adjustment, suggest that the projected 18 percent increase of the carbon dioxide concentration by the year 2000 (to about 379 ppm) would result in an increase of the surface temperature of about one-half degree and a stratospheric cooling of $0.5°$ to $1°C$ at 20 to 25 km; a doubling of the carbon dioxide concentration over the present level would result in an increase of the surface temperature of about $2°C$, and a $2°$ to $4°C$ decrease in the stratosphere at the same level. We would like to emphasize, however, that these computations neglect the important interacting dynamics and thermodynamics of the atmosphere, as well as the ocean-atmosphere interaction. This neglect makes the computed temperature changes very uncertain.

Recommendations

1.   We recommend that comprehensive global numerical-dynamical models, including ocean-atmosphere interaction and cloud variation, be developed and applied to the study of expected circulation, precipitation, and temperature patterns for the levels of CO_2 anticipated for the future.

2.   We recommend that the effects of stratospheric cooling be studied by developing models that permit cloud formation at these heights.

1.4.3

Effects of Changing Particles and Turbidity

All atmospheric particles in the size range 0.1- to $2.5\text{-}\mu$ radius are of special interest to the earth's heat budget, since they are relatively abundant in the atmosphere and are well recognized as having the greatest effect in scattering, absorbing, and attenuating solar radiation. Particles in this size range also may have an effect on terrestrial long-wave radiation, which is usually ignored. Larger particles that might interfere more strongly with long-wave radiation are less abundant in most natural aerosols, but may become important locally around industrial pollution sources or in heavily dust-laden air (Bryson and Baerreis, 1967).

It has commonly been assumed that the attenuation of direct solar radiation by particles is mainly attributable to scattering rather than to absorption (McCormick and Ludwig, 1967). For particles in the 0.1- to 2.5-μ radius range, scattering is predominantly in forward (downbeam) directions, with only a small fraction of the total (approximately 10 percent) in the backward (upbeam) directions. Any increase of backscatter tends to increase the albedo (reflectivity) of the atmospheric layer containing the particles. Whether increased backscatter increases or decreases the albedo of the combined atmosphere and underlying surface, however, depends upon the surface albedo.

Recently the neglect of *absorption* in atmospheric particulate layers has been criticized (Charlson and Pilat, 1969; Robinson, 1970). There are indications that in certain man-made aerosols, especially those of industrial origin containing carbon, iron oxides, and some other materials, the amount of absorption is of the same order as the amount of backscattering (Roach, 1961; Waldram, 1945). Backscattering of solar radiation by itself reduces the total heating within and below the scattering medium, since the fraction scattered back to space is not available for heating. Absorption, on the other hand, increases the heating within the absorbing medium. The net effect of the two processes on heating within and below an aerosol layer depends on their relative magnitudes. To determine this net effect in actual cases, it is considered necessary to measure the absorption and backscattering coefficients by the use of airborne instruments (Roach, 1961; Robinson, 1966).

Recent calculations by M. Atwater (1970) and by J. M. Mitchell (1970a) of the combined effects of absorption and scattering by aerosols and reflection from the surface indicate that a warming effect is attainable with relatively small (and nearly equal) absorption and backscattering.

It should be stressed that the situation with regard to stratospheric aerosols, or to high-tropospheric aerosol layers, is different from that of low- and middle-tropospheric aerosol layers in one very important respect. The heating caused by absorption within the former group of aerosol layers occurs well above the earth's surface. In such a case, surface air is likely to be cooled regardless of the magnitude of the absorption (and associated warming) in the aerosol layer itself. Direct observation of very small changes of atmospheric surface temperature following volcanic eruptions which hurled volcanic particles into the stratosphere (Mitchell, 1970b) tends to confirm this. (There is a further discussion of this in Section 1.4.7, where the Agung eruption is compared to the potential SST particulate addition.)

Another quite different effect of aerosols, not related to their optical properties, is their role as condensation and freezing nuclei. Condensation nuclei, which are effective in forming water droplets at low supersaturation, are fairly well understood, but freezing nuclei are not. In particular, the wide fluctuations of freezing nuclei in space and time and their association with some human activities strongly suggest that precipitation from cold clouds may be inadvertently influenced by man. (Cloud seeding with silver iodide is a case of purposeful modification, of course.) This could have an effect on a regional basis (Schaefer, 1969), but in view of the limited lifetimes of freezing nuclei in the atmosphere it seems highly unlikely that man will be able to influence the global freezing nuclei content.

Conclusions

1. Tropospheric particles can produce changes in
   a. albedo;
   b. cloud reflectivity;
   c. solar radiation reaching the surface and the upward flux of terrestrial infrared radiation;
   d. cloud formation and precipitation.

While the specific effects on climate cannot be designated at this time, it appears clear that any such effects are enhanced as the number of atmospheric particles is increased. However, the nature of the impact on global climate by changes in the population of tropospheric particles remains undetermined in both sign and magnitude.

2. Large variations in the stratospheric population of particles may produce correspondingly large fluctuations in stratospheric temperatures. For example, in the case of the Agung eruption, an

increase of 6° to 7°C was found in the equatorial stratosphere. However, only a small decrease is apparent in the global surface temperature following volcanic eruptions.

3.   The area of greatest uncertainty in connection with the effects of aerosols on the heat balance of the atmosphere lies in our current lack of knowledge of their optical properties, in both the solar and terrestrial parts of the spectrum (approximately 0.3 to 30 μ).

Recommendations

1.   We recommend that *in situ* scattering and absorption measurements be made under a variety of locales and conditions and that the radiative effects of particles be investigated.

2.   We recommend that laboratory and field research be conducted in many areas, including

a. sources, transport and dispersal, and purging mechanisms for contaminant particles;

b. nature and statistics of particle populations in the troposphere and the stratosphere;

c. particle production and transformation in the atmosphere;

d. particle effects on the physics of clouds.

The various tropospheric contamination patterns of particles should be identified and characterized by kind of contaminant population, geographic extent, and typical time histories downwind (in urban areas, coastal zones, deserts, and so on).

3.   We recommend that the effects of particles on radiative transfer be studied, and that the results be incorporated in mathematical models to determine the influence of particles on planetary circulation patterns.

1.4.4

Atmospheric Clouds

Cloud Distribution

The distribution of cloudiness is an important atmospheric parameter because of its central role in affecting the radiation budget of the atmosphere and, therefore, the evolution of climate.

Clouds reflect solar radiation (the reflectivity differs according to cloud type) and in general act as efficient radiators for terrestrial radiation. Also, because cloud transmissivity of solar radiation is higher than cloud transmissivity of terrestrial (infrared) radiation, clouds are important contributors to the so-called "greenhouse" effect.

As just indicated, clouds produce two opposite effects. Increased cloudiness results in an increased albedo which reduces the solar energy available to heat the surface but, at the same time, increases the back (downward) radiation to raise the surface temperature. It has been shown (Ohring and Mariano, 1964; Manabe and Wetherald, 1967) that the *net* result of increased low or middle cloudiness is to lower the surface temperature. Increased cirrus clouds, on the other hand, would warm the surface if the clouds were efficient (blackbody) radiators, or have practically no effect if the clouds transmitted infrared radiation. Some observations indicate that thin cirrus has an infrared transmissivity of about 95 percent, but thick cirrus (of the order of 5 km) may have a transmissivity of as little as 50 percent (see, for instance, Kuhn and Weickmann, 1969).

Before 1960 our knowledge of the cloudiness distribution was derived almost entirely from land and ship observations. Most of the observations were made over populated land areas and over narrow, but frequently used, shipping lanes. Relatively few observations were available in the tropics and in the Southern Hemisphere in general.

The ground-based observations indicate that the earth is covered about 52 percent by clouds with possible global variations from about 50 to 54 percent. These observations also suggest large seasonal variations in the tropics and subpolar latitudes and marked longitude variations at all latitudes. There are, however, no reliable data of the year-to-year variation of total global cloudiness, although the results of some studies have suggested a recent (the last ten years) increase in cirrus clouds. (See Section 1.2.4.)

Since 1960, satellite observations have been used to give some indications of the global cloudiness distribution. Because of the resolution and video systems used (there is a need in this area to set a proper threshold brightness for cloud detection), individual, small cloud groups are generally smeared out, and often thin cirrus clouds are not detected (although cirrus clouds could be inferred from infrared sensors). Horizontal spatial resolution of the order of 500 meters is needed therefore in order to get "correct" cloud population counts. We understand that this is within the capabilities of satellite systems now in operation and of those planned for the near future. If the brightness contrast is low, the threshold will be set at a low value and satellite-observed clouds may be reported in higher amounts than actually present. Some simultaneous observations by different satellite systems indicate that the problem of threshold brightness cannot be neglected if correct cloud populations are to be reported. Also, it is sometimes difficult to distinguish clouds from highly reflecting surface systems such as ice fields or desert areas. Despite these limitations, satellite observations have already shown the existence of unique cloud cluster formations and large cloudiness variability in the tropics and have provided considerable information on relative cloudiness in the Southern Hemisphere, where there was almost no prior data.

Since 1966, daily (and with the launch of ITOS-1 in 1970, twice daily) global cloud cover maps have been available *on a routine basis*. Detailed ground-based cloud cover data should be used to calibrate the satellite observations so the inferences made from satellite information in sparse data areas will be correct. This coordination will help in standardization of the satellite observations so that the absolute value of the total cloud cover and its variations can be determined.

Clouds appear in the atmosphere in different forms and at different levels. This information is, of course, exceedingly important in heat budget and climatic change studies. However, there is no practical method known, at present, for determining precisely the distribution of cloud amounts by type and height. Models of these distributions have, therefore, been developed, based on the regular synoptic, and aircraft, meteorological reports. It would be highly desirable if surface and satellite systems (using visible, infrared, and microwave sensors) could be coordinated to derive the necessary picture of the three-dimensional cloud patterns.

A global system of standardized cloud observations would provide the information for studies of the month-to-month, year-to-year and trend variations, as well as latitudinal and hemispheric differences, in cloudiness and cloudiness patterns (whether natural or man-made). It would be helpful if all this information were processed in a form so that it would be readily available for improved radiation studies and for the analysis of climatic trends, both "natural" and inadvertent.

Optical Properties of Clouds

Clouds reflect, absorb, and transmit electromagnetic radiation, but these properties depend very sensitively on the spectral range considered; on the solar zenith angle (for visible radiation); and on the cloud parameters such as the cloud's type, height, and thickness, droplet size and water content, and particulate concentration (that is, so-called clean or dirty clouds), and the nature of the surface below the cloud. There are many observations of optical characteristics and some theoretical studies of (ideal) clouds. The average cloud albedo varies from about 10 to 25 percent for cirrus to about 70 to 85 percent for thick cumulus clouds (Sellers, 1965; Kondratyev, 1969). The absorptivity of these clouds, although generally small (2 to 3 percent), can apparently be as high as about 30 percent (Robinson, 1970). The large measured variations in the transmission of solar radiation by clouds (differences of a factor of 2, in some cases for the same cloud type) cannot at present be explained.

Some specific optical properties that need to be studied, and their importance in environmental studies, are outlined below:

VISIBLE SPECTRUM

1.   Reflectively as a function of solar zenith angle, cloud parameters and, in the case of nonstratoform clouds, of the cloud array. Observations show that the reflectivity could vary from cloud to cloud within the same cloud type. Some of this variation might be due to the difference between a single cloud and the combined reflection of a group of clouds (for example, cumuluform) of the same type. This variation could be as high as 20 percent reflectivity. In view of the possible increase of cirrus clouds in the atmosphere, its reflective characteristic must be determined more accurately than is known at present.

2.   Absorption. The existing data (both theoretical and observational) give, for instance, absorptivities in cumulus clouds varying from 6 to 25 percent. This information needs to be verified and to be defined better as a function of the cloud parameters. The

absorption of solar radiation in clouds can have a pronounced effect on the heating of the atmosphere.

3.   The contamination of clouds by nonwater substances could change the reflective and absorptive properties of the cloud. It is likely that "dirty" clouds have lower albedos and higher absorptivities than "clean" clouds. So-called dirty clouds are found in areas of high particulate concentrations.

INFRARED SPECTRUM. The radiative characteristics of clouds are known somewhat better for infrared radiation than for visible. The most important problem here is determining the emissivity ("blackness") of the various clouds as a function of thickness and composition (water or ice). In particular, it is necessary to determine the degree of blackness of cirrus clouds, both natural and those produced, for instance, by jet aircraft. The radiative properties of these clouds affect their detection by satellites but also could have some influence on temperature variation in the upper atmosphere. Although preliminary studies do not indicate that cirrus clouds play a large role in determining the surface temperature, the possible consequence of cloudiness variations for the earth's climate is very profound and requires further careful study.

Recommendations

1.   We recommend that there be global observations of cloud distribution and its temporal variations. The importance of cloudiness in the atmosphere stems from its relatively high albedo for solar radiation and its central role in the various processes involved in the heat budget of the earth-atmosphere system. High spatial resolution satellite observations (of the order of 500 m) are required to give "correct" cloud population counts and to establish the existence of long-term trends in cloudiness (if there are any).

2.   We recommend as being of fundamental importance studies of the optical (visible and infrared) properties of clouds as functions of the various relevant cloud and impinging radiation parameters. These studies, particularly for cirrus, should include the effect of particles on the reflectivity of clouds and a determination of the infrared "blackness" of clouds.

1.4.5

Effects of Heat Released

By the year 2000, the thermal power output of the world is estimated (Greenfield, 1970) to be of the order of 3×10^{13} watts, which is about 6 times the present output. Distributed over the globe, this is an insignificantly small amount. However, over cities the heating will be of the order of 1 percent of the solar constant, or about 10 percent of the solar heat absorbed at the ground. If this pattern continues, this will produce heat islands (cities or larger areas that are warmer than their surroundings) not much different in temperature elevation from those existing now, though more extensive in area (Peterson, 1969).

Recommendation

We recommend that the possibility of wider implications of the heat released be tested by numerical modeling. Specifically, the potential regional effects of aggregated heat islands should be examined (see the discussion of urbanization in Section 1.4.6).

1.4.6

Effects of Changing the Earth's Surface

Introduction

Some of the most important changes to the earth's surface are being produced by increasing urbanization, increasing land cultivation, and new water bodies. In addition, human activity may produce climatic changes that will affect the snow cover, thus producing a feedback.

Urbanization

Urbanization increases the area of cities or of city conglomerations. The heat island effect (the excess of temperature over the surroundings) is due largely to man-made heat sources in urban areas (Peterson, 1969). The temperature excess should remain of about the same magnitude as it is now; but the elevated temperatures should cover larger areas. Probably, the effects of such increased heat islands will be at least regional. For example, if the city borders an ocean or lake, the usual sea breeze would be suppressed by the diurnal variation of the heat island (that is, the city is relatively warm at night). The heat island effect causes air to rise over a city. This leads to a slightly larger precipitation probability over a city than over its surroundings.

Cities are rougher than most other regions. Hence, there is more turbulence and stronger vertical mixing over cities than elsewhere. On the other hand, the winds tend to be somewhat weaker.

Air pollution in cities, which is created by industry, incineration, automobiles, and space heating, causes snow to become dirty and to melt faster than it does in the cities' surroundings. The melting is further aided by the "heat island." Particles created over the city can travel large distances, dirtying snow in large areas and causing rapid melting. In this way, urbanization may affect the wintertime albedo in a large area. Normally, however, the albedo of cities is not substantially different from that of their surroundings.

Effects of Water Bodies

Artificial lakes, produced by dams, have had negligible effects on the general climate and even on the shower activity downwind from them. This has occurred because the lakes have been created in otherwise quite dry areas. It is difficult to see how conventional damming operations now contemplated could affect the global climate.

However, certain more ambitious and speculative operations have been proposed. For example, if the Bering Straits were dammed and the Arctic Ocean were ice-free, it is possible that land glaciers would form in Canada and spread into the United States (if, as Ewing and Donn suggested in 1956, the principal requirement for another ice age is the supply of water made available by an ice-free Arctic Ocean). Similarly, damming the Congo would affect the climate at least over a large portion of central Africa. The anticipated area is large enough to suggest that global changes of climate as a result of such projects cannot be completely ruled out.

Effect of Changes of Agricultural Patterns

On a global scale, changes of agricultural patterns will affect only a few percent of the earth's surface. New cropland will be produced, roughly half from grassland and half from forests. The change from grass to crops has very little effect on albedo and other surface features. Change from forest to crops increases the albedo (particularly when snow covered), decreases the water available to the atmosphere (unless irrigated), and decreases the roughness. It is difficult to foresee any changes in global climate as a result of these factors.

Effect of Oil Spills

It has been suggested that oil spilled on the oceans may form extensive monomolecular layers which will prevent evaporation. Observations with the kind of oil spilled at sea have shown, however, that the molecular structure of this oil is sufficiently irregular to permit evaporation (LaMer, 1962). In fact, evaporation is quite vigorous over the Mediterranean, into which oil is being spilled extensively. Also, the oil layer is usually perturbed by the various motions at the sea surface. As a result, evaporation is reduced only in the immediate region of the original spills, and such regions will not cover a significant area of the sea by the year 2000.

Also, oil spills appear cooler than their surroundings in the infrared spectrum and warmer in the microwave region (see, for example, Hanst, Lehmann, and Reichle, 1970). Again, however, these temperature effects are small, even on a local basis.

Ice and Snow

If climatic change, man-made or natural, lowers the temperatures in arctic regions, ice sheets will spread. Since snow and ice have higher albedos than any other ground cover, the result would be a further reduction of temperature, resulting in a further increase in the ice. This behavior suggests an instability which, unless checked by other factors, could lead to an equator-to-pole ice cover with an initial mean annual temperature drop of only about 2°C (Budyko, 1968, 1969, 1970; Sellers, 1969, 1970).

We know that even at the peak of the Pleistocene ice ages ice did not penetrate beyond middle latitudes. Hence factors not considered, such as cloudiness, the limited moisture supply, the oceans, and atmospheric motions counteract this "albedo instability." Nevertheless, it is an important process that must be included in theories of climatic change of any kind.

Conclusions

Surface changes which contribute to the production of CO_2 or particles are discussed elsewhere (Sections 1.2.1 and 1.2.2 of this report and 2.6 of report of Work Group 2).

Important among the remaining surface changes are, first, man's activities that modulate snow and ice cover, particularly in polar regions; and, second, some possible projects involving the production of new, very large water bodies.

Over and above these projects, only increased urbanization is of possible global importance, as it produces extended areas of contiguous cities. Still, it is not certain whether effects of urbanization extend far beyond the general region occupied by the cities.

Recommendation

We recommend that, before actions are taken which result in some of the very extensive surface changes described, mathematical models be constructed which simulate their effects on the climate of a region or, possibly, of the earth.

1.4.7

Effects of Jet Aircraft

There are two quite distinct subjects that we must deal with here: (1) the question of the effects of condensation trails (contrails) formed by current, commercial, subsonic, jet aircraft flying in the upper troposphere, and occasionally in the lowest part of the stratosphere (below about 12 km, or 40,000 feet); and (2) the question of the possible future effects of the contamination by SSTs of the higher stratosphere (at about 20 km, or 65,000 feet).

Effects of Cirrus from Jet Contrails

Added contrails or cirrus clouds can affect the climate through both the heat balance and the nucleating role of falling ice crystals.

If an increase in cirrus cloudiness were to become significant as a result of jet activity in the upper atmosphere, the most important global effect would be an increase in the earth-atmosphere albedo. Less solar energy would be available to drive the atmosphere; however, as in all such suggestions, we must caution that until a realistic simulation of the process is available we cannot be sure that the atmosphere might not compensate for the artificial increase in cloudiness. There have been measurements and theoretical estimates of the immediate, static effects of added cirrus clouds on the temperature near the ground. The answers depend sensitively on the reflectivity, emissivity, and temperature of the clouds. In fact, one may reverse the sign of the temperature change by changes of the emissivity, for example.

It is known that falling ice crystals can nucleate supercooled water clouds into which they fall. The result would be a quicker initiation of the precipitation process than would otherwise occur. Much more work in this field would be needed before the importance of such artificial seeding effects can be evaluated.

Conclusion

Two weather effects from enhanced cirrus cloudiness are possible. First, the radiation balance may be slightly upset and, second, cloud seeding by falling ice crystals might initiate precipitation sooner than it would otherwise occur.

Recommendations

1.   We recommend that changes in cirrus cloud population be monitored directly through the program recommended in Section 1.4.4.

2.   We recommend that the radiative properties of representative contrails and contrail-produced cirrus clouds be determined.

3.   We recommend that the significance, if any, of ice crystals falling from contrail clouds as a source of freezing nuclei for lower clouds be determined.

Effects of SST Contaminants in the Stratosphere

Two products of the SST are of concern: water vapor and particles. The increased water vapor in the stratosphere may have three consequences: (1) a "greenhouse" radiation effect, (2) ozone depletion as suggested by Hampson (1964, 1966) or (3) cloud formation.

1.   The greenhouse radiation effect has been calculated by Manabe and Wetherald (1967) for a quasi-static atmosphere and is reproduced as Figure 1.6. The increase in ground level temperature, derived from Figure 1.6, is less than $0.10°$ C for the world average increase of 0.2 ppm of water vapor by mass.

2.   The inclusion of hydrogen (water vapor) in photochemical reactions will produce a smaller equilibrium ozone concentration than will a dry atmosphere. The reduced ozone would then admit more ultraviolet radiation to the lower atmosphere, a potentially

Figure 1.6 Surface Air Temperature Change Due to Increased Stratospheric Water Vapor, with Average Cloudiness and Corrective-Radiative Equilibrium

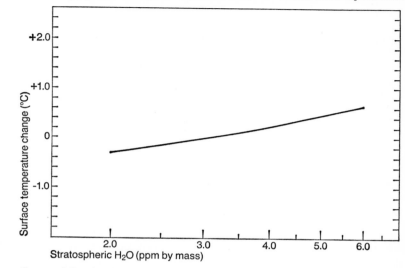

Source: Manabe, 1970, after Manabe and Wetherald, 1967

undesirable condition for man and the environment. The quantitative aspects of the inclusion of moisture have been treated by London and Park (1970) and are illustrated by Figure 1.7. With the best present-day reaction rates and solar insolation amounts, equilibrium ozone distributions were calculated and the results are shown in Figure 1.7; one with no water vapor, curve 1; a second with 5 ppm (5×10^{-6}), curve 2; and a third with 20 ppm (2×10^{-5}), curve 3. It will be noted that water vapor alters the ozone concentration mainly above 50 km. Since the amount of ozone above 50 km is very small compared with that below, major changes at the higher altitudes are of little significance in changing the total ozone in the entire column which controls the ground receipt of ultraviolet radiation. It must be noted that water vapor increases—either from 3 to 5 ppm (the likely maximum) or from 3 to 3.2 ppm (the world average)—fall between the curves, thus producing very small changes in total ozone. Interpolating, one finds a decrease of about 0.02 cm of ozone (standard temperature and pressure [STP]) for an increase in stratospheric moisture of 2 ppm, and 0.002 cm for the average

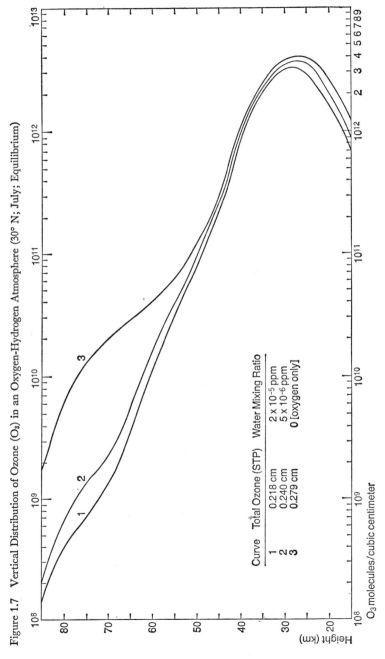

Figure 1.7   Vertical Distribution of Ozone (O₃) in an Oxygen-Hydrogen Atmosphere (30° N; July; Equilibrium)

Curve	Total Ozone (STP)	Water Mixing Ratio
1	0.218 cm	2 x 10⁻⁵ ppm
2	0.240 cm	5 x 10⁻⁶ ppm
3	0.279 cm	0 [oxygen only]

O₃ molecules/cubic centimeter

Height (km)

Source:: London and Park, 1970

global increase in stratospheric moisture of 0.2 ppm. Both of these changes in total ozone lie well within the routine, day-to-day, and geographical, variability.

These calculations must, however, be considered provisional. The photochemical reaction rates are, in some instances, imperfectly known; and in these calculations there is no adjustment allowed for transport or for the depletion of water vapor by photochemistry (which may reduce the new equilibrium water vapor concentration and consequently also reduce its influence).

3.   Nacreous, or mother-of-pearl, clouds occur in the height range 20 to 30 km and are most often seen over Norway in winter when the sun is below the horizon by several degrees. A condition for their occurrence seems to be a temperature near $-80°C$ or lower. Their lifetimes are not known with certainty, as the conditions for good visibility do not last long (World Meteorological Organization, 1956; Hesstvedt, 1959).

In the Southern Hemisphere over the Antarctic, a thin cloud veil in the lower stratosphere is often observed from midwinter to the beginning of October. The cloud frequently appears to cover the whole sky (Liljequist, 1956). Clouds are also observed near 80 km in summer at high latitudes (Ludlam, 1957).

The low temperature associated with nacreous clouds above the tropopause, coupled with their seasonal and geographical variations, lends support to the hypothesis that they are produced when local saturation occurs. At 24 km (30 mb), for example, when a mixing ratio of 3×10^{-6}g/g is assumed, the frost point is $-87°C$. It is quite conceivable that this temperature can be reached on individual days, particularly if the local vertical velocity is reinforced by mountain waves, as many people have suggested.

Suggestions concerning modification of high-level clouds from the injection of water vapor by the SST are based on the idea that a higher mixing ratio would permit clouds to form because the frost point could be reached at a higher temperature and, therefore, over a wider geographical and altitude range.

Recently Mastenbrook (1970) showed that the humidity mixing ratio over the Naval Research Laboratory's Chesapeake Bay Station east of Washington, D.C., had increased from about 2 ppm (by mass) in 1964 to 3 ppm in 1970. The upward trend is continuing and has no evident explanation. There are no other measurements adequate to confirm or reject Mastenbrook's findings or to indicate its geographic extent. A continued upward trend would clearly increase the likelihood of both natural cloudiness and artificial clouds from SST operations.

A further argument favoring cloudiness formation lies in the predictions of stratospheric cooling by the increase of atmospheric CO_2 (irrespective of the source of the CO_2). Manabe and Wetherald's calculation (1967) suggests that a tenfold greater cooling in the stratosphere than heating of the lower atmosphere would be caused by an increase in the CO_2 "greenhouse" effect. By 1985–1990 this could amount to several degrees C cooling of the stratosphere from the projected CO_2 increase due to the burning of fossil fuels by all sources (not only the SST).

The effects of SST particles (soot, converted hydrocarbons, and converted SO_2) can be assessed by analogy with the Agung dust cloud, in which nature may be said to have performed a most helpful experiment for us. The temperature in the equatorial stratosphere following the Agung eruption rose 6° to 7°C, and remained at 2° to 3°C above its pre-Agung level for several years, perhaps due to other volcanic activity in the tropics that has maintained a relatively high particle content (see Figure 1.8). Thus, something like a thirtyfold increase in stratospheric particles made itself felt in ordinary radiosonde temperature observations. The effect of Agung was observed for about 15 degrees of latitude, north and south of the source, at stations around the world. Such data have been analyzed, for example, from Australian stations 027 (Lae 6.5°S, 147°E), 120 (Darwin 12°S, 131°E), 461 (Giles 25°S, 128°E), 865 (Laverton 38°S, 145°E), and New Zealand station 780 (Christchurch 43°S, 173°E) (Newell, 1970a).

Although the analogy between the insertions of particles into the stratosphere by Agung and by the SSTs is tempting, there are several rather basic differences between the two situations. For example, the particles from Agung must have spread through a vertically much thicker layer, whereas the SST exhaust will all be injected in the limited range of its flight altitudes. It should also be noted that Agung is located near the equator. There, the temperature is lower than to the south or north; and it also increases upward from 17 km to 40 km. The SSTs, on the other

Figure 1.8   Temperatures at Different Pressures above Port Hedland, Australia

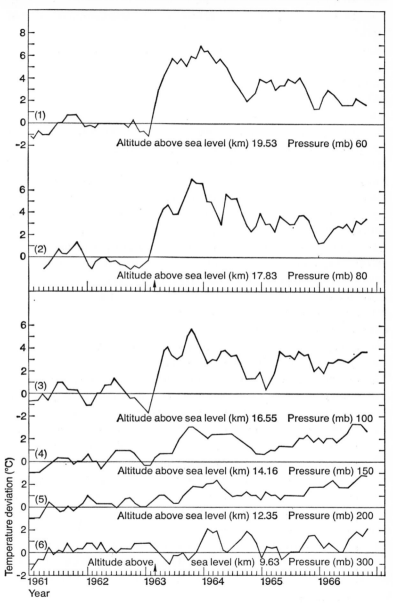

Source: Newell, 1970a

Note: Arrow at March 1963 notes the Mt. Agung eruption.

hand, are likely to produce most of their particles at middle and high latitudes. In winter the temperature at 55°N at 20 km is actually larger than either to the south or to the north, and varies little in the vertical from 17 to 30 km. In summer the pole is warmer than other latitudes, but horizontal temperature gradients are weak.

For these and other reasons, the effects produced by Agung and by the SSTs may be quite different, even if the number of particles introduced is the same.

Our calculations show that the peak particulate loadings from the SST operations may be the same as the post-Agung observed concentrations for sulfates, and if one accepts the smaller pre-Agung values (which are less certain), then almost all the present estimates of particle concentrations from SST activities exceed the natural background concentrations. These increased particle loadings may raise the temperature and play an important role in the stratospheric heat balance, as did the Agung sulfate cloud.

Conclusions

1. The added SST water vapor in the stratosphere may introduce the following three effects (in order of likely importance):

   a. Stratospheric clouds, already observed in the polar night, may increase in frequency, thickness, and extent. The effect of SST water vapor will be heightened by the increasing trends in CO_2 and natural stratospheric water vapor (observed only over Washington, D.C.).

   b. Direct radiation effects will, according to Manabe and Wetherald's quasi-static radiation calculations, result in warming of air at ground level in regions of peak moisture concentration by less than a tenth of a degree C on a world-wide basis and cooling in the stratosphere by a few degrees C. (The actual global effect would be smaller than the expected changes due to CO_2 increases.)

   c. The reduction of ozone due to water vapor interaction (in a static photochemical model) has been estimated to lie well within the present day-to-day and geographical variability of total ozone.

2.   The direct role of quantities of CO, CO_2, NO, NO_2, SO_2, and hydrocarbons in altering the heat budget is small. It is also unlikely that their involvement in ozone photochemistry is as significant as water vapor.

3.   The SO_2, NO_x, and hydrocarbons can undergo complex reactions that produce particles. Increased particulate loadings may raise the temperature and play an important role in the stratospheric heat balance.

Recommendations

A feeling of genuine concern has emerged as a result of these facts and conclusions concerning the SST operations; we perceive that man's activities as he flies the projected 500 SSTs can have a clearly measurable effect in large regions of the world where they will fly, and quite possibly on a global scale. The effects will be most pronounced in the stratosphere, but we cannot exclude the possibility of significant effects at the surface. We must emphasize that, due to the uncertainties in the available information and its interpretation, we cannot be certain about the magnitude of the various effects.

Therefore, we recommend that uncertainties about SST contamination and its effects be resolved before large-scale operation of SSTs is implemented.

Specifically, we recommend the following program of action:

1.   Begin now to monitor the lower stratosphere for water vapor, cloudiness, oxides of nitrogen and sulfur, hydrocarbons, and particles (including the latter's composition and size distribution).

2.   Determine whether additional cloudiness or persistent contrails will occur in the stratosphere as a result of SST operations particularly in cold areas, *and* the consequences of such changes.

3.   Obtain better estimates of contaminant emissions, especially those leading to particles, under simulated flight conditions and under real flight conditions, at the earliest opportunity.

4.   Using the data obtained in carrying out these recommendations, estimate the change in particle concentration in the stratosphere attributable to future SSTs and its impact on weather and climate.

References

American Chemical Society (ACS), 1969. *Cleaning Our Environment: The Chemical Basis for Action* (Washington, D.C.: ACS).

Appleman, H., 1965. Investigation of the effect of contrails on cirrus frequency and coverage (unpublished).

Arrhenius, S., 1896. The influence of the carbonic acid in the air upon the temperature of the ground, *Philosophical Magazine, 41*.

Arrhenius, S., 1903. *Lehrbuch der kosmichen Physik 2* (Leipzig: Hirzel).

Atwater, M. A., 1970. Investigation of the radiation balance for polluted layers of the urban environment (unpublished).

Bainbridge, A. E., 1970. Data on the secular increase of atmospheric carbon dioxide at Mauna Loa, 1963–1969 (private communication to L. Machta).

Bhandari, N., Lal, D., and Rama, 1966. Stratospheric circulation studies based on natural and artificial radioactive tracer elements, *Tellus, 18*.

Bolin, B., and Bischof, W., 1969. Variations of the carbon dioxide content of the atmosphere (Stockholm: University of Stockholm, Institute of Meteorology), Report AC-2.

Brookhaven National Laboratory (BNL) 1969. *The Atmospheric Diagnostic Program at Brookhaven National Laboratory: Second Status Report, November 1969*. (Long Island, New York: BNL). Report #BNL 50206 (T-553).

Brown, C. W., and Keeling, C. D., 1965. The concentration of atmospheric carbon dioxide in Antarctica, *Journal of Geophysical Research, 70*.

Bryson, R. A., 1968. All other factors being constant, *Weatherwise, 21*.

Bryson, R. A., and Baerris, D. A., 1967. Possibilities of major climate modification and their implications: Northwest India, a case study, *Bulletin of the American Meteorological Society, 48*.

Budyko, M. I., 1968. On the origin of glacial epochs, *Meteorologiia i Gidrologiia, 11*.

Budyko, M. I., 1969. The effect of solar radiation variations on the climate of the Earth, *Tellus, 21*.

Budyko, M. I., 1970. Comments on A global climatic model based on the energy balance of the earth-atmosphere system, *Journal of Applied Meteorology, 9*.

Cadle, R. D., Bleck, R., Shedlovsky, J. P., Blifford, I. A., Rosinski, J., and Lazrus, A. L., 1969. Trace constituents in the vicinity of jet streams, *Journal of Applied Meteorology, 8*.

Cadle, R. D., Lazrus, A. L., Pollock, W. H., and Shedlovsky, J. P., 1970. The chemical composition of aerosol particles in the tropical stratosphere, *Proceedings of the American Meteorological Society Symposium on Tropical Meteorology* (unpublished).

Chamberlain, T. C., 1899. An attempt to frame a working hypothesis of the cause of glacial periods on an atmospheric basis, *Journal of Geology, 7*.

Charlson, R. J., and Pilat, M. J., 1969. Climate: the influence of aerosols, *Journal of Applied Meteorology, 8*.

Cobb, W. E., and Wells, H. J., 1970. The electrical conductivity of oceanic air and its correlation to global atmospheric pollution, *Journal of Atmospheric Science*, forthcoming.

Cotten, J., 1970. Analysis and third degree polynomial fit of carbon dioxide concentration data from Mauna Loa (private communication to L. Machta).

Dopplick, T. G., 1970. Radiative Heating of the Tropical Atmosphere, in *The General Circulation of the Tropical Atmosphere* (Cambridge: The M.I.T. Press), forthcoming.

Ewing, M., and Donn, W. L., 1956. A theory of ice ages, *Science, 123*.

Federal Aviation Administration (FAA), 1967. *The U.S. SST, a Report on Economic Feasibility* (Washington, D.C.: FAA).

Fletcher, J. O., 1969. Controlling the planet's climate, in *Impact of Science on Society, 19*.

Flohn, H., 1969. *Climate and Weather*, B. V. deG. Walden, trans. (New York: McGraw-Hill).

Gebhart, R., 1967. On the significance of the shortwave CO_2 absorption in investigations concerning the CO_2 theory of climatic change, *Archiv für Meteorologie, Geophysik und Bioklimatologie, B 15*.

Greenfield, S. M., 1970. Projection and distribution of waste thermal energy, background paper prepared for SCEP (unpublished).

Hampson, J., 1964. Photochemical behaviour of the ozone layer, *Canadian Armament Research Division Establishment Technical Note #1627* (Val Cartier, Quebec: C.A.R.D.E.).

Hampson, J., 1966. Chemiluminescent emissions observed in the stratosphere and mesosphere, *Les Problèmes Météorologiques de la Stratosphère et de la Mésosphère* edited by Centre National d'Etudes Spatiales (Paris: Presses Universitaires de France).

Hanst, P., Lehmann, J., and Reichle, H., 1970. Remote detection of air and water pollution from satellites and/or airplanes, background paper prepared for SCEP (unpublished).

Hess, S. L., 1959. *Introduction to Theoretical Meteorology* (New York: Henry Holt and Company).

Hession, J. P., 1970 (private communication to A. K. Forney).

Hesstvedt, E., 1959. Mother of pearl clouds in Norway, *Geophysica Norvegica, 20*.

Johnson, F. S., 1969. Origin of planetary atmospheres, *Space Science Reviews, 9*.

Junge, C. E., 1963. *Air Chemistry and Radioactivity* (New York: Academic Press).

Keeling, C. D., 1970. Atmospheric carbon dioxide: Long-term trends, background paper prepared for SCEP (unpublished).

Kelley, J. J., Jr., 1969. An analysis of carbon dioxide in the arctic atmosphere near Barrow, Alaska, 1961–1967 (University of Washington), Report #NR 307-252.

Kondratyev, K. Ya., 1969. *Radiation in the Atmosphere* (New York and London: Academic Press).

Kuhn, P. M., and Weickmann, H. K., 1969. High altitude radiometric measurements of cirrus, *Journal of Applied Meteorology, 8*.

Lamb, H. H., 1966. *The Changing Climate* (London: Methuen Books).

LaMer, V. K. (editor), 1962. *Retardation of Evaporation by Mono-Layers: Transport Processes* (New York: Academic Press).

Lees, L., 1970. Waste heat in the Los Angeles Basin (private communication to S. M. Greenfield).

Leipunskii, O. I., Konstantinov, J. E., Fedorov, G. A., and Scotnikova, O. G., 1970. Mean residence time of radioactive aerosols in the upper layers of the atmosphere based on fallout of high altitude tracers, *Journal of Geophysical Research, 75*.

Liljequist, G. H., 1956. *Norwegian-British-Swedish Antarctic Expedition, 1949–1952, Scientific Results,* Vol. II (Oslo: Norsk Polarinstitutt).

London, J., and Park, J., 1970. Ozone photochemistry in a wet atmosphere, forthcoming.

London, J., and Sasamori, T., 1970. Radiative energy budget of the atmosphere, *Space Research,* forthcoming.

Lorenz, E. N., 1970. Climatic change as a mathematical problem, *Journal of Applied Meteorology, 9.*

Ludlam, F. H., 1957. Noctilucent clouds, *Tellus, 9.*

Ludwig, J. H., Morgan, G. B., and McMullen, T. B., 1969. Trends in Urban Air Quality (paper presented at the American Geophysical Union meeting, San Francisco).

McCormick, R. A., and Ludwig, J. H., 1967. Climate modification by atmospheric aerosols, *Science, 156.*

Machta, L., and Carpenter, T., 1970. Secular Growth of Cirrus Cloudiness, forthcoming.

Machta, L., and Hughes, E., 1970. Atmospheric oxygen in 1967 to 1970, *Science, 168.*

Manabe, S., 1970. Surface air temperature change due to increased stratospheric water vapor, with average cloudiness and convection—radiative equilibrium (private communication to L. Machta).

Manabe, S., and Bryan, K., 1969. Climate calculations with a combined ocean-atmosphere model, *Journal of Atmospheric Science, 26.*

Manabe, S., and Wetherald, R. T., 1967. Thermal equilibrium of the atmosphere with a given distribution of relative humidity, *Journal of Atmospheric Science, 24.*

Martell, E. A., 1970. *Transport Patterns and Residence Times for Atmospheric Trace Constituents vs. Altitude,* Advances in Chemistry Series, 93 (Washington, D.C.: American Chemical Society).

Mastenbrook, H. J., 1970. Concurrent measurements of water vapor and ozone over Washington, D.C., during 1969 and 1970 (private communication to L. Machta).

Mitchell, J. M., Jr., 1965. Theoretical paleoclimatology, *The Quaternary of the United States* edited by H. E. Wright, Jr. and D. G. Frey (Princeton: Princeton University Press).

Mitchell, J. M., Jr., 1968. Causes of climatic change, *Meteorological Monographs, 8,* No. 30 (Lancaster: Lancaster Press, Inc.).

Mitchell, J. M., Jr., 1970a. The effect of atmospheric particulates on radiation and temperature, background paper prepared for SCEP (unpublished).

Mitchell, J. M., Jr., 1970b. A preliminary evaluation of atmospheric pollution as a cause of the global temperature fluctuation of the past century, *Global Effects of Environmental Pollution,* edited by J. F. Singer (Dordrecht: Reidel Publishing Company), forthcoming.

Möller, F., 1963. On the influence of changes in the CO_2 concentration in air on the radiation balance at the Earth's surface and on the climate, *Journal of Geophysical Research, 68.*

Newell, R. E., 1970a. Stratospheric temperature change from the Mount Agung volcanic eruption of 1963, *Journal of Atmospheric Science,* forthcoming.

Newell, R. E., 1970b. Modification of stratospheric properties by trace constituent changes, *Nature, 227.*

Newell, R. E., and Dopplick, T. G., 1970. The effect of changing CO_2 concentration on radioative heating rates (unpublished).

Ohring, G., and Mariano, J., 1964. Changes in the amount of cloudiness and the average surface temperature of the Earth, *Journal of Atmospheric Science, 21.*

Pales, J. C., and Keeling, C. D., 1965. The concentration of atmospheric carbon dioxide in Hawaii, *Journal of Geophysical Research, 70.*

Peterson, J. T., 1969. *The Climate of Cities: A Survey of Recent Literature* (Washington, D.C.: National Air Pollution Control Administration, U.S. Government Printing Office).

Plass, G., 1956. The carbon dioxide theory of climatic change, *Tellus, 8.*

President's Science Advisory Committee (PSAC), 1965. *Restoring the Quality of Our Environment* (Washington, D.C.: U.S. Government Printing Office).

Raschke, E., and Bandeen, W. R., 1970. The radiation balance of the planet Earth from radiation measurements of the satellite Nimbus II, *Journal of Applied Meteorology, 9.*

Roach, W. T., 1961. Some aircraft observations of fluxes of solar radiation in the atmosphere, *Quarterly Journal of the Royal Meteorological Society, 87.*

Robinson, G. D., 1966. Some determinations of atmospheric absorption by measurement of solar radiation from aircraft and at the surface, *Quarterly Journal of the Royal Meteorological Society, 92.*

Robinson, G. D., 1970. Some meteorological aspects of radiation and radiation measurement, *Advances in Geophysics,* edited by H. E. Landsberg and J. Van Mieghem (London: Academic Press).

Schaefer, V. J., 1966. Ice nuclei from automobile exhaust and iodine vapor, *Science, 154.*

Schaefer, V. J., 1969. The inadvertent modification of the atmosphere by air pollution, *Bulletin of the American Meteorological Society, 50.*

Sellers, E. D., 1965. *Physical Climatology* (Chicago: University of Chicago Press).

Sellers, E. D., 1969. A global climatic model based on the energy balance of the earth-atmosphere system, *Journal of Applied Meteorology, 8.*

Sellers, E. D., 1970. Reply to comments of M. I. Budyko, *Journal of Applied Meteorology, 9.*

Spirtas, R., and Levin, H. J., 1970. *Characteristics of Particulate Patterns* (Washington, D.C.: National Air Pollution Control Administration, U.S. Government Printing Office).

Suess, H. E., 1965. Secular variations of the cosmic ray produced carbon 14 in the atmosphere and their interpretations, *Journal of Geophysical Research, 70.*

Swihart, J., 1970. Results of survey by the Boeing Company of sulfur content of present jet fuels (private communication to W. W. Kellogg).

Telegadas, K., and List, R. J., 1964. Global history of the 1958 nuclear debris and its meteorological implications, *Journal of Geophysical Research, 69*.

Thompson, J., 1970. Information on the calculated emissions of the GE-4 engine (private communication to L. Machta).

Tyndall, J., 1863. On Radiation Through the Earth's Atmosphere, *Philosophical Magazine* (4th series), *25*.

United Nations, 1956. World Energy Requirements, in *Proceedings of the International Conference on the Peaceful Uses of Atomic Energy*, Vol. I (New York: United Nations Department of Economic and Social Affairs, United Nations).

United Nations. *World Energy Supplies*, Statistical Papers, Series J (New York: Statistical Office, Department of Economic and Social Affairs, United Nations).

Waldram, J. M., 1945. Measurements of the photometric properties of the atmosphere, *Quarterly Journal of the Royal Meteorological Society, 71*.

World Meteorological Organization, 1956. *International Cloud Atlas* (Geneva: Atar S. A.).

2.
Work Group on Ecological Effects

Chairman
Frederick E. Smith
HARVARD UNIVERSITY

Geirmunder Arnason
CENTER FOR THE ENVIRONMENT
AND MAN, INC.

Edward Goldberg
SCRIPPS INSTITUTION OF OCEAN-
OGRAPHY

M. Grant Gross
STATE UNIVERSITY OF NEW YORK

Arthur Hasler
UNIVERSITY OF WISCONSIN

Bruce B. Hanshaw
U.S. GEOLOGICAL SURVEY

J. B. Hilmon
U.S. FOREST SERVICE

Dale W. Jenkins
THE SMITHSONIAN INSTITUTION

Philip C. Kearny
AGRICULTURAL RESEARCH SERVICE

Frank Lowman
PUERTO RICO NUCLEAR CENTER

Jerry S. Olson
OAK RIDGE NATIONAL LABORA-
TORY

Joseph Reid
SCRIPPS INSTITUTION OF OCEAN-
OGRAPHY

Clarence Tarzwell
FEDERAL WATER QUALITY ADMIN-
ISTRATION

Edward Wenk
UNIVERSITY OF WASHINGTON

George Woodwell
BROOKHAVEN NATIONAL LABORA-
TORY

Rapporteurs
Stephen Burbank
HARVARD LAW SCHOOL

Robert Stoller
HARVARD LAW SCHOOL

2.1
Introduction

This report focuses on three major areas of ecological concern in the biosphere:* (1) man's overall effect on terrestrial vegetation and terrestrial ecosystems; (2) his cumulative effect on the estuaries and coastal oceans that support most of the world's fisheries, and (3) problems relating to the biology of the oceans. In all three areas a lack of critical information severely handicapped problem evaluation. Nevertheless, major environmental problems have been identified that exist now or seem certain to exist soon.

Only a small part of the biosphere is managed by man, yet he relies upon all of it for the regulation of climate, composition of the atmosphere, supplies of clean water, and many other services. In this sense the entire biosphere is used. The increasing scope and intensity of such use, compounded by practices whose side effects have spread throughout the globe, threaten the capacity of the biosphere to continue its normal functions. Man may suddenly be faced with far greater management obligations than he is willing or able to undertake. Changes in human action are urgently needed, both to ameliorate our impact upon the environment and to develop new management capabilities.

2.2
The Problem

The paragraphs that follow summarize information on man's total impact, general changes in ecosystems, and the consequences of such changes for man. These are the general arguments in support of the contention that relations between man and biosphere are approaching a stage of crisis.

2.2.1
Present Scale of Human Activity

Man is now using about 41 percent of the total land surface (Table 2.1). More than half of the remainder is not usable because it is too cold, frozen, or mountainous. The most optimistic report on land use that has appeared in recent times (President's Science Advisory Committee [PSAC] Report, 1967) estimates that as much as two-thirds of the total land surface is usable. These

* The biosphere of the earth is composed of all life-forms and of that portion of air, land, and water in which life exists. The following is the scientific definition of biosphere to which this Work Group subscribed: a zone of the earth intersecting with the lower atmosphere, hydrosphere, and upper lithosphere.

Table 2.1
Present and Potential Uses of the Land Surface of the Planet
(Percent of total area)

Use	Present	Potential
Croplands	11	24
Rangelands	20	28
Managed forests	10	15
Reserves (80% forest)	26	0
Not usable	33	33
Total Land	100	100

Source: President's Science Advisory
Committee (PSAC), 1967.

figures can be taken as the upper limit. If an estimate is weighted
for the different intensities of different land uses, we appear to
be using about half of the earth's land resources at the present
time (Table 2.1).

The Food and Agriculture Organization of the United Nations [FAO] projects very little exploitation of the remaining potential (Table 2.2). The land already in use tends to be the best
for such use, and intensification of management will probably be
chosen rather than expansion into marginal lands. Expansion has
ended in the developed nations and, by 1985, is expected to be
ended in most of the world. Thus, except for an expansion in the
utilization of forest land, use is expected to become stabilized.
The difference between the FAO projections and the potential
estimated in the PSAC Report is only a factor of 2, which is not

Table 2.2
Past and Projected World Land-Use Changes
(Percent of total land area)

Use	1950–1968	1968–2000
Cropland	+1.5	+0.4
Rangeland	+5.2	−1.3
Total forest	−0.2	−1.3

Sources: Food and Agriculture Organization (FAO), 1951, 1958, 1969.

large in comparison with problems of population growth over the next century.

The present scale of human activity can also be estimated for a number of specific man-induced processes in comparison with natural rates of geological and ecological processes. H. J. M. Bowen has compared global mining activities with estimates of the annual discharge of materials through rivers into the oceans. Such natural rates are well suited to an analysis of pollution problems in the coastal oceans (Bowen, 1966). Twelve cases for which the man-induced rate is as large or larger than the natural rate are shown in Table 2.3. With a 5 percent annual growth increment in the mining industries, many more materials will soon fall into this category (Table 2.4). Although data on some natural rates may be erroneous and therefore lead to some incorrect inclusions and omissions, the table nonetheless gives a reasonable overview.

Table 2.3
Man-Induced Rates of Mobilization of Materials Which Exceed Geological Rates As Estimated in Annual River Discharge to the Oceans
(Thousands of metric tons per year)

Element	Geological Rates[a] (In Rivers)	Man-Induced Rates[b] (Mining)
Iron	25,000	319,000
Nitrogen	8,500	9,800 (consumption)
Manganese	440	1,600
Copper	375	4,460
Zinc	370	3,930
Nickel	300	358
Lead	180	2,330
Phosphorus	180	6,500 (consumption)
Molybdenum	13	57
Silver	5	7
Mercury	3	7
Tin	1.5	166
Antimony	1.3	40

Sources:
[a] Bowen, 1966.
[b] United Nations, *Statistical Yearbook*, 1967. Data for mining except where noted.

Table 2.4
**World Average Annual Rates of Increase for the Period 1951–1966
for Selected Aspects of Human Activity[a]**

	Percentage[b]
Agricultural production	3
Industry based on farm products	6
Mineral production (including fuel)	5
Industry based on mineral products	9
Construction and transportation	6
Commerce	5

Source: Digested from United Nations, *Statistical Yearbook,* 1967.

[a] Data are for the world excluding Mainland China.
[b] Rates are in constant dollars.

These comparisons show that at least some of our actions are large enough to alter the distribution of materials in the biosphere. Whether these changes are problems depends upon the toxicity of the material, its distribution in space and time, and its persistence in ecological systems. Better estimates of natural processes, and of man-induced rates that include waste contributions from other industries, would be very useful.

If total production rates of materials are small in comparison with natural rates, problems may be locally intensive but are not expected to be large on a global scale. Where technological exploitation or production rates are high, data on loss rates to the environment are needed. In the absence of such data it is impossible to ascertain the effects of technology on the environment. For example, an industry with a high recycling rate and an accumulating inventory in products may have a small output of waste.

2.2.2
Growth in Human Demands

The world population has been increasing at about 2 percent per year, a rate that produces a doubling every 35 years (United Nations, *Statistical Yearbook,* 1967). Population alone, however, does not reflect our total demand upon the environment. As our numbers increase, the technology needed to support people must increase even more. This arises in part from the increasing complexity of activities associated with larger populations and in part from the increasing difficulty of achieving higher and higher rates of production of goods. The average rates of annual increase of relevant activities for the last fifteen years are shown in Table 2.4.

The problem of achieving higher rates of production is well defined in agriculture, where increasing yields are obtained only from still greater inputs. Table 2.5 shows these increases over a fifteen-year period.

Finally, Table 2.6, from the FAO *Production Yearbook* for 1963, shows how the quantity of pesticide application relates to increasing food supplies. Note that Japan, with twice the U.S. yield of food per acre, uses ten times as much pesticide, while we, with just over twice the yield of Africa, use eleven times as much pesticide.

Such a high ecological cost for increased food production may be the result of a deeply entrenched agricultural system, or it may be unavoidable. In either case it poses a very difficult specific problem associated with population growth.

A measure of *ecological demand* is needed that summarizes man's impact upon the environment. This would include both the utilization of resources and the disposal of wastes and would reflect the intensity of interaction between the biosphere and the aggregate of the man-made world. One such statistic is the Gross Domestic Product (minus services) which is compiled by the

Table 2.5
World Average Rates of Increase for the Period 1951–1966 for Selected Aspects of Human Activity Related to Food Production

	Percentage[a]
Food	34
Tractors	63
Phosphates	75
Nitrates	146
Pesticides	300

Source: Digested from United Nations, *Statistical Yearbook*, 1967.

[a] Rates in constant dollars.

Table 2.6
Pesticide Usage and Agricultural Yields in Selected World Areas

Area or Nation	Pesticide Use		Yield	
	Grams per hectare	Rank	Kilograms per hectare	Rank
Japan	10,790	1	5,480	1
Europe	1,870	2	3,430	2
United States	1,490	3	2,600	3
Latin America	220	4	1,970	4
Oceania	198	5	1,570	5
India	149	6	820	7
Africa	127	7	1,210	6

Source: FAO, *Production Yearbook*,
1963.

United Nations (Table 3 in United Nations, *Statistical Yearbook*, 1968). Since 1950 this index has been increasing between 5 and 6 percent per year, doubling in 13½ years, which appears to be a reasonable estimate for the current rate of increase in ecological demand. If this should continue, then, in the time taken for the human population to double, our total ecological demand would increase sixfold.

More than anything else, such a rapid rate of growth suggests why environmental problems seem to have erupted so suddenly, why the future will surely bring more problems than the present, and why slowness to respond may be disastrous.
2.2.3
Global Ecological Effects
The significant aspect of human action is man's total impact on ecological systems, not the particular contributions that arise from specific pollutants. Interaction among pollutants is more often present than absent. Furthermore, the total effect of a large number of minor pollutants may be as great as that of one major pollutant. Thus, the total pollution burden may be impossible to estimate except by direct observation of its overall effect on ecosystems.

Recognition of this overall effect is essential for defining the real pollution problem. However, the effects and ultimate costs of specific pollutants must also be documented if effective control is to become possible. Of the pollutants that pose threats to ecological systems, this Work Group chose to concentrate on four classes of materials released by man into the environment. Case studies for these are presented in the next section of this Work Group report.

Selective Impairment of Predators

Among animals, predators (those that eat other animals) are generally more sensitive to environmental stress than herbivores (those that eat plants). Furthermore, in aquatic systems the top-level predators (which eat other predators, for example, most game fish) are the most sensitive of all. This has been found for stressors such as oxygen deficiency, thermal stress, toxic materials, and pesticides. It is as general in terrestrial as in aquatic systems, and among insects and other invertebrates as well as among vertebrates.

Examples of this phenomenon are numerous. Overenrichment by sewage waste and fertilizer runoff of freshwaters, or pollution with industrial wastes, leads to the rapid loss of trout, salmon, pike, and bass. Spraying crops for insect pests has inadvertently killed off many predaceous mites, resulting in outbreaks of herbivorous mites that obviously suffered less. Forest spraying has similarly "released" populations of scale insects after heavy damage to their wasp enemies.

In addition to suffering direct damage from pollutants, predators also suffer from the damage done lower in the food web to the herbivores and plants. In effect such pollutants compete with the predators, their damage resulting in a loss of food to the predators. Both in nature and in computer models, severe damage to lower levels in the food chain usually leads to the extinction of the predator before its food is extinct.

Persistent pesticides reveal a third mechanism by which predators may suffer more than their prey. Such fat-soluble pesticides as DDT are concentrated as they pass from one feeding level to the next. In the course of digestion a predator retains rather than eliminates the DDT content of its prey. The more it eats, the more DDT it accumulates. The process results in especially high concentrations of toxins in predacious terrestrial vertebrates.

Excessive Damage at One Brief Stage in Life Cycles

The need for more thorough toxicity studies is presented later in this report. The frequency with which damage is brief but intense, hitting one particular stage without necessarily harming older or younger stages, makes cause-and-effect studies very difficult. Usually some phase of the reproductive cycle is involved. Oyster larvae are wiped out by pollution with copper that poses no danger to established oysters. DDT can kill fish larvae at the moment of yolk absorption without seeming to have harmed the parent that put the pesticide into the egg. The final effect is that entire populations can disappear, and the apparent health of the older individuals may lead to a lack of corrective action until it is too late.

This cause of local extinction is less likely to be selective among herbivores and predators than the causes of damage already discussed.

Greater Instability in Ecosystem Regulation

Terrestrial ecosystems of landscape size usually contain hundreds of species of plants and thousands of species of animals, but the herbivores rarely consume more than 5 to 15 percent of the vegetation.

Occasionally the set of processes that generally keeps herbivores in check fails, and they become numerous enough to damage the foliage appreciably. Even so, it is rare for such outbreaks to be severe. Ecosystem regulation is so general that exceptions are objects of intense study. The tundra-lemming system in the arctic and the conifer-spruce budworm system in northern forests appear to be two examples of systems in which strong fluctuations are inherent (Morris and Miller, 1954; Pitelka, Tomick, and Treichel, 1955).

The normal degree of control is best summarized by pointing out that for each species of terrestrial plant there are about 100 species of animals capable of eating it, yet most of the time, most of the plant production falls to the ground uneaten. For many insects able to defoliate trees the population density rarely rises as high as one per tree.

Nevertheless, this stability is fragile enough to be easily destroyed. The list of disturbances that lead to population outbreaks (almost always of herbivores) includes almost anything that removes a number of species or impairs the health or numbers of predators. Forests are familiar with the consequences that may follow cleaning out underbrush, selectively removing tree species, or growing single-aged one-species stands (see, for example, Davis, 1954).

The two general effects discussed strongly affect the frequency and severity of population explosions in ecosystems. The loss of species from excessive damage leaves a system less able to function or to recover from dysfunction, while selective damage to predators can precipitate population outbreaks.

Attrition of Ecosystems

The most general effect of pollutants on ecosystems combines all of the effects discussed previously with additional direct depression of plant or animal vigor. This leads to overall deterioration of the system, characterized by instability and species loss. The result is a system of few species, generally weeds, pollution-tolerant in the sense of being able to support high death rates. The only further step is annihilation, a stage that so far has occurred only in areas of extreme pollution.

Many lakes and urban centers have severely deteriorated ecosystems. Less severe deteriorations occur more commonly, often as temporary afflictions in ecosystems that otherwise manage to survive intact. Once an ecosystem is severely damaged and becomes unattractive, its death is usually considered an improvement by the people who live within it.

This general problem is labeled "attrition" because it lacks discrete steps of change. Stability is lost more and more frequently, noxious organisms become more common, and the aesthetic aspects of waters and countryside become less pleasing. This process has already occurred many times in local areas. If it were to happen gradually on a global scale, it might be much less noticeable, since there would be no surrounding healthy ecosystems against which to measure such slow change. Each succeeding generation would accept the status quo as "natural."

The gradual decline in ecosystem function brings with it a decline in services for man. When at the same time nature becomes unappealing to man, the decision is easily reached to dispense with the natural systems (for example, filling in a polluted lake). The real costs of dispensing with more and more natural functions need to be appraised.

2.2.4
Environmental Services

It is a mark of our time, and a signal of the degree to which man is ecologically disconnected, that the benefits of nature need to be enumerated. More important, however, is the need to evaluate each service in terms of the cost of replacing it or the cost of doing without it (including future costs that may result from the loss of additional services). Such an evaluation cannot be performed here, but a listing of a few benefits with some indication of their value can be attempted.

Pest Control

At least 99 percent of the potential plant pests in the world are controlled by natural means. Man attempts to control only a few thousand, most of them on crops (Hoffman and Henderson, 1967). The total use of pesticides has not yet been sufficient to destroy the numerous herbivore-predator systems that operate so well in natural vegetation, although the list of crops that have lost their natural means of control increases steadily.

Forest growth is slowed by pest outbreaks, a cost that has long been appreciated in forestry, leading to major research programs on forest pests. The loss of value in parks and suburban areas is primarily one of aesthetics, perhaps best estimated by the price that people are willing to pay for a tree. If the pest problem is ignored and annual defoliation accepted as a norm, the ecosystem will respond by producing a scrubby kind of vegetation dominated by annual weeds.

Insect Pollination

A few groups of insects pollinate most of the vegetables, fruits, berries, and flowers. The unintended killing of pollinators with insecticides has often occurred on a local scale, usually noticed only when a desired crop fails to set. Farmers have learned to be careful.

The problem arises anew in the context of general vegetation rather than particular crops. The current rate of increase in the use of insecticides, the large amount of acreage involved, and the slowly increasing broad-scale applications to large areas combine as a threat to the continued function of insect pollinators.

   The cost of pollination under human management is so great that the production of dependent crops would decrease greatly. Most wildflowers and whole sections of natural flora would disappear. Except for a few favored crops, man is certainly unable to replace this service; the cost of irreversible loss will have to be placed against the apparent advantage of any massive insecticide program.

Fisheries

Although aquaculture produces important sources of fish protein, most of the harvest comes from fisheries where production is entirely natural. Marine fishery yields have been increasing at 5 percent per year. Although basic productivity may remain high in lakes that are overenriched (Lake Erie is still the most productive of the Great Lakes), the kinds of fish caught are usually the least desirable on the market.

   The present threat to marine fisheries lies not so much in the danger of polluting the grounds where the fish are caught but in the pollution of inshore areas where many species reproduce. These nurseries are being dredged, filled, and polluted with both nutrients and heavy metals. The value of commercial yields becomes a cost to be considered in evaluating policy for estuaries and coastal oceans.

Climate Regulation

Natural vegetation air-conditions the landscape and also serves as windbreaks. The day-night regimes of temperature, humidity, and wind on barren land indicate the role of vegetation in the human environment. If the human population expands to the degree that the attrition of ecosystems leads to poor vegetation, this free service will not be well performed. The cost of air-conditioning the landscape or the cost of living with an unpleasant or even unbearable outdoor climate will ultimately have to be considered.

Soil Retention

Vegetation holds the soil on the land. Regulating the total volume of our soils, which decrease slowly in croplands and increase slowly under vegetation, is another asset provided free in nature.

Flood Control

Flood frequency and intensity vary greatly with the presence of vegetation along waterways. Vegetation distributes the flow of water evenly. By contrast, flash floods and mudflows are common in the barren gullies of the southwest deserts.

Soil Formation

Soil is a mixture of plant debris and weathered rock, formed through the joint action of bacteria, fungi, worms, soil mites, and insects. These agents comprise about 40 percent of the total biomass of terrestrial animals. Without these biological actions, plant debris disintegrates very slowly, and the soil fails to develop water-absorbent, ion-exchange properties. Larger insects also play a vital role in the processes of decay that recycle materials by breaking down branches and tree trunks.

Cycling of Matter

The cycling of the elements in organisms depends upon green plants as the basic energy source, combined action of animals and microbes for decomposition, and specific groups of plants for several specialized steps involving nitrogen and sulfur transfer. Without these three major components, material would accumulate in one part of the cycle and ecosystems would deteriorate.

Composition of the Atmosphere

Much of the oxygen, carbon dioxide, and nitrogen in the air and dissolved in the oceans passes through organisms in ecological cycles. These cycles, interacting with the geological cycle of erosion and deposition, lead to the regulation of atmospheric composition on a geological time scale. If natural life were largely destroyed, these functions could be taken over by regulating industries. However, we would have to balance all of our own actions, for example, reducing carbon dioxide as fast as it is made in combustion.

The amounts of nitrate in land and water, and the amounts of methane and ammonia in the air and water, are regulated primarily by microbes. The loss of this function would coincide with the ultimate destruction of life.

2.2.5

Evaluation of Current Status

Man does not yet threaten to annihilate natural life on this planet. Nevertheless, his present actions have a considerable impact on ecosystems, and his future actions and numbers will certainly have even more. The critical issue is the danger that we may curtail an environmental service without being able to carry the loss or we may irreversibly lose a service that we cannot live comfortably without.

In general, the expected losses from present impacts do not exceed our capacity to carry the burden; this leads us to the conclusion that an intractable crisis does not now seem to exist. Our growth rate, however, is frightening. The impact of two, four, or eight times the present ecological demand will certainly incur greater losses in the environment. If the process of change were gradual, the present ecological advantage that is reflected in our 5 to 6 percent annual growth would taper off in the face of decreased environmental services, and growth would be correspondingly slowed. Instead, the risk is very great that we shall overshoot in our environmental demands (as some ecologists claim we have already done), leading to cumulative collapse of our civilization.

It seems obvious that before the end of the century we must accomplish basic changes in our relations with ourselves and with nature. If this is to be done, we must begin now. A change system with a time lag of ten years can be disastrously ineffectual in a growth system that doubles in less than fifteen years.

2.3
Case Studies

DDT, mercury, petroleum, and phosphorus were selected for case studies, not because they are necessarily the four most important pollutants but because they differ significantly in sources, routes, and effects. Considered together, they represent a broad spectrum of the problems and issues involved when assessing man's impact on the biosphere.

All of these cases deal primarily with the marine environment. They provide examples of what we do and do not know about specific pollutants. As noted earlier, however, such examples should ultimately be considered with respect to the total pollution burden in terrestrial and aquatic ecosystems.

2.3.1
DDT in the Marine Environment*
The Ecological Impact

The acute and chronic toxicity of DDT has been identified by

* This case study is derived from a report on pesticides in the marine environment prepared at the Study by the following SCEP participants and

observing its effects under controlled laboratory conditions. The exposure of test populations of marine fauna to serial dilutions of DDT in flowing seawater has shown that it affects growth, reproduction, and mortality at concentrations currently existing in the coastal environments. These laboratory effects and their field counterparts may be summarized as follows.

PLANKTON. In the open ocean phytoplankton are at the base of the food chain and may act as primary concentrators of DDT from the water. Laboratory evidence is available demonstrating its inhibitory effect on photosynthesis in single-celled marine plants (Wurster, 1968; Menzel, Anderson, and Randtke, 1970a). It is doubtful that these results are ecologically meaningful. The concentrations necessary to induce significant inhibition exceed expected concentrations in the open ocean and exceed by ten times its solubility (1 ppb) in water (Bowman, Acree, and Corbett, 1960). However, toxicity may vary interspecifically and, if not universally toxic, may exert some control on species succession in the near-shore environment.

If DDT is concentrated in surface oil film, it is not improbable that concentrations there may reach levels sufficient to cause acute toxicity to plants. Considering that this layer may extend to 1 mm in depth, its effect on total producion within the euphotic zone would be about 10^{-6} (100 m depth). It should be borne in mind, however, that plants and/or bacteria may provide an effective means of extracting DDT from the surface film.

FISH. Marine fish are almost universally contaminated with DDT residues. There is an expected concentration of such residues in tissues high in lipid content, such as the ovary. In the speckled sea trout on the south Texas coast, DDT residues in the ripe eggs are about 8 ppm. This level may be compared with the residue of 5 ppm in freshwater trout, which causes 100 percent failure in the development of sac fry or young fish. There is presumptive evidence for similar reproductive failure in the sea trout. Sea trout inventories in the Laguna Madre in Texas have shown a

members of a Workshop of the National Academy of Sciences Committee on Oceanogrephy (NASCO), who were SCEP consultants: E. D. Goldberg, Chairman;† P. Butler;† G. Ewing; M. Ingham; D. Jenkins; P. Kearney; B. Ketchum; P. Meier;† D. Menzel;† J. Reid; R. Risebrough;† L. Stickel;† G. Woodwell; S. Burbank, Rapporteur.

† Indicates NASCO Workshop members. E. D. Goldberg was a full-time SCEP participant and also Chairman of the NASCO group.

progressive decline from 30 fish per acre in 1964 to 0.2 fish per acre in 1969. It is significant that no juvenile fish have been observed there in recent years, although in less-contaminated estuaries 100 miles away, there is a normal distribution of sea trout year classes (Butler, 1969).

Declines in the productivity of fish in California coastal waters have not yet been documented as being correlated with pesticide residues. The harvest of California mackerel has been halted, however, because DDT residues exceed FDA guidelines even in the processed product.*

Laboratory experiments have also established the concentrations of several chlorinated hydrocarbons, including DDE†, that damage reproductive success of birds, fish, and marine invertebrates. Concentrations of DDT in species from the marine environment exceed those found to have deleterious effects in the laboratory and have been correlated with population decreases or reproductive failures of a number of marine species. Signs of incipient damage that can be expected to develop with continuing accumulation have also been reported.

Experimental evidence from the calanoid copepod shows that the development of adults from nauplear stages is completely blocked when hatched from egg-bearing females maintained in seawater containing 10 parts per trillion DDT (Menzel, Anderson, and Randtke, 1970b). Significant mortality was observed at 5 parts per trillion. These concentrations of DDT are lower than expected in rainwater falling on the sea surface (80 parts per trillion [Tarrant and Tatton, 1968]). Data are not now available to extrapolate from these observations to the open oceans, since no measurements of DDT concentrations are available from these waters.

CRUSTACEANS. Bioassay tests show that laboratory populations of commercial species of shrimp and crabs as well as zooplankton are killed by exposure to DDT, in the parts per billion range

* The 5 ppm guideline for DDT in the edible portion of fish was set by the FDA so that daily consumption of $1/4$ pound of fish, plus the dietary intake from other foods (data derived from FDA total's diet studies), would not exceed the FAO-WHO acceptable daily intake for DDT. FDA determines when safe levels are exceeded by monitoring foods and products transported interstate. The agency takes legal action to seize unsafe products, through the Federal Courts.
† DDE is a metabolite of DDT.

(Butler, 1964). Continuous exposure of shrimp to DDT concentrations of less than 0.2 ppb caused 100 percent mortality in less than 20 days (Nimmo, Wilson, and Blackman, forthcoming). DDT concentrations of this magnitude have been detected in Texas river waters flowing into commercially important shrimp nursery areas (Manigold and Schulze, 1969). We can be certain that significant mortalities of juvenile crustaceans are increasing in such contaminated areas. In California the declining production of Dungeness crabs is associated with observed DDT residues in the developing larvae.

MOLLUSKS. DDT characteristically interferes with the growth of oysters at levels as low as 0.1 ppb (Butler, 1966) in the ambient water. Mollusks generally concentrate chlorinated pesticides and thus serve as indicators of pollution levels in marine waters. Coastal monitoring samples have demonstrated that the magnitude of DDT residues in mollusks is directly correlated with its application rates in adjacent river basins (Butler, 1967).

BIRDS. In experimental studies, DDE in the diet resulted in thin eggshells and reduced hatching success of mallard ducks; p,p'-DDT (an isomer of DDT) produced the same effects, but to a lesser degree (Heath, Spann, and Kreitzen, 1969).

Deaths of bald eagles, common loon, peregrine falcons (Jefferies and Prestt, 1966), and sea eagles from coastal localities have been correlated with lethal amounts of DDT in body tissues.

In the marine ecosystems of southern California, where concentrations of the DDT compounds in fish may exceed 10 parts per million (Risebrough et al., forthcoming), two species, the bald eagle and the peregrine falcon, have disappeared from the wilderness environment of the Channel Islands (Herman, Kirven, and Risebrough, 1970). In this area the brown pelicans and double-crested cormorants are no longer able to reproduce. Only one brown pelican was known to hatch in southern California in 1969 for about 1,200 nesting attempts (Risebrough, Sibley, and Kirven, forthcoming). The cause in each case was a failure of the eggs to hatch because of breakage during incubation. Concentrations of DDE, polychlorinated biphenyls (PCB), and the other chlorinated hydrocarbons in the lipid fractions of the eggs were correlated with shell thickness and compared with concentrations of these compounds in brown pelican eggs from the Isles Coronados,

Mexico, the Isles San Martin and San Benitos, Baja California, Mexico, the Gulf of California, four Florida localities, Jamaica, Venezuela, Panama, and Peru. Shell thickness changes were evident in the Mexican and Florida eggs, and all ranges of shell thickness were observed, from normal to an almost complete absence of calcium carbonate (Jehl, forthcoming; Risebrough and Anderson, forthcoming; Schreiber, forthcoming).

Below a critical shell thickness, amounting to approximately a 21 percent reduction, the eggs invariably broke. Analysis of approximately 180 eggs and egg fragments showed that thickness decreases with increasing DDE content.

Shell thinning has also been detected in marine and coastal birds in northern California, including the ashy petrel (Risebrough and Coulter, unpublished manuscript). Continued buildup of DDT in this ecosystem and in other marine ecosystems around the world will cause reproductive failures in these and other marine species.

Biochemical Effects

Several physiological effects of DDT and its residues account for shell thinning and for the abnormal behavior observed in contaminated populations. In affecting nerves, chlorinated hydrocarbons, including DDE, are believed to block the ion transport process by inhibiting ATPase (an enzyme, otherwise called myosin) in the nerve membrane that causes the required energy to be made available (Matsumura and Patil, 1969). Transport of ionic calcium across membranes such as those in the shell gland of birds is also an energy-requiring process dependent upon membrane ATPase (Skou, 1965). Inhibition of these enzymes by DDE could account for the concentration-effect curves obtained for shell thickness and DDE concentration in eggs of the brown pelican, double-crested cormorant (Anderson et al., 1969) and herring gull (Hickey and Anderson, 1968). DDE has also been found to inhibit the enzyme carbonic anhydrase (Bitman, Cecil, and Fries, 1970; Peakall, 1970; Risebrough, Davis, and Anderson, forthcoming), essential for the deposition of calcium carbonate in the eggshell and for the maintenance of pH gradients across membranes, such as those in the shell gland. Inhibition of this enzyme by such drugs as sulfanilamide results in the production of thin-shelled eggs.

The chlorinated hydrocarbons, including DDE, induce mixed-function oxidase enzymes in the livers of birds and mammals that hydroxylate and render water-soluble foreign, lipid-soluble compounds (Conney, 1967; Risebrough et al., 1968). Induction is usually temporary, ending when the inducing materials are themselves metabolized. DDE is comparatively resistant to degradation by the induced enzymes, so that they may persist as inducers without being degraded. The induced enzymes may therefore become constitutive.

The steroid hormones such as estrogen and testosterone (Conney, 1967; Peakall, 1970) and thyroxine (Schwartz et al., 1969) are metabolized at higher rates when these enzymes are induced. Lower estrogen concentrations are present in pigeons fed p,p'-DDT (Peakall, 1970). Birds may also show symptoms of hyperthyroidism when fed a chlorinated hydrocarbon (Jefferies, 1969; Jefferies and French, 1969). An increasing number of instances of abnormal behavior are being reported in contaminated populations, including herring gulls and the brown pelicans (Gress, forthcoming). These abnormalities also affect reproductive success and most likely result from hormone imbalance caused by the activity of the nonspecific enzymes induced by the chlorinated hydrocarbons.

U.S. and World Production of DDT

The U.S. production figures for DDT are presented in Table 2.7. The United States now utilizes less than 30 percent of the DDT manufactured (see Table 5.2). For DDT, we have chosen to use an integrated world production figure of 2.0×10^6 and a yearly production of 10^5 metric tons in calculations. It should be noted clearly that there are insufficient data available to determine the actual world production figures of DDT.

Transport of DDT Residues to the Marine Environment*

The two principal routes for transport of DDT residues from places of application on land to the ocean are rivers and the atmosphere.

SURFACE RUNOFF. Total annual surface runoff for all continents has been estimated at 3.8×10^{13} metric tons (Nace, 1967). The maximum concentration of DDT residues in water reported in a

---

* DDT residues is a term defining DDT, DDE, and DDD. DDD and DDE are metabolites of DDT, with the former being a pesticide in its own right.

Table 2.7
Production of DDT in Units of 10^3 Metric Tons/Year
(United States only)

Year	DDT	Year	DDT
1944	4.4	1957	56.6
1945	15.1	1958	66.0
1946	20.7	1959	71.2
1947	22.5	1969	74.6
1948	9.2	1961	77.9
1949	17.2	1962	75.9
1950	35.5	1963	81.3
1951	48.2	1964	56.2
1952	45.4	1965	64.0
1953	38.4	1966	64.2
1954	44.2	1967	47.0
1955	59.0	1968	63.4
1956	62.6	Total	1,220.0

Source: *Chemical Economics Handbook*,
1969.

survey of rivers of the western United States was about 100 parts
per trillion (Bailey and Hannum, 1967; Manigold and Schulze,
1969). If all rivers of the world contain this maximum, 3.8×10^3
metric tons of DDT residues would be transported annually to
the sea. This total would be approximately 3 percent of the
annual production of the world. It is, however, almost certainly
high, perhaps by a factor of 10. DDT residues transported by the
Mississippi River into the Gulf of Mexico have been estimated
to be 10 metric tons annually (Risebrough et al., 1968). The total
river output from the continental United States might be twice
that of the Mississippi, approaching 20 metric tons. Domestic U.S.
use of DDT between 1961 and 1968 has ranged between 5 and
8×10^4 metric tons per year, thousands of times more than that
carried in the rivers. It seems unlikely that much more than
1/1000 of the annual production of DDT (100 metric tons) could
reach the oceans by surface runoff.

ATMOSPHERIC TRANSPORT. DDT residues enter the atmosphere by
several routes, including aerial drift during application, by rapid
vaporization from water surfaces (Acree, Berova, and Bowman,

1963) and by vaporization from plants and soils (Nash and Beal, 1970). Once in the atmosphere it may travel great distances, entering the sea in precipitation or in dry fallout. There are few data for estimating the rates of transfer. The most extensive sampling of DDT residues in precipitation has been in Great Britain, where total accumulation was measured at 7 stations between August 1, 1966, and July 1967. The mean concentration for all rainwater samples was 80 parts per trillion (Tarrant and Tatton, 1968). This concentration is about twice that reported for meltwaters of recent antarctic snow (Peterle, 1969). The DDT residues in South Florida precipitation averaged 1,000 parts per trillion in 18 samples taken at 4 sites between June 1968 and May 1969 (Yates, Holswade, and Higer, 1970).

Total annual precipitation over the oceans has been estimated as 2.97×10^{14} metric tons (Sverdrup, Johnson, and Fleming, 1942). If this contained an average of 80 parts per trillion, a total of 2.4×10^4 metric tons of DDT residues would be transported annually to the oceans. This is about one-quarter of the estimated total annual production of DDT. Thus, the atmosphere appears to be the major route for transfer of DDT residues into the oceans.

Distribution of DDT in the Marine Environment

Few data are available to document the concentration of DDT in the open ocean environment. Gray and sperm whales contain up to 0.4 and 6 ppm DDT residues in their blubber respectively (Wolman and Wilson, 1970). The former feed largely on benthic organisms in the Chukchi and Bering Seas, and the latter on larger pelagic organisms. Sea birds (petrels and shearwaters), which feed on planktonic organisms far from land, have concentrations of DDT residues as high as 10 ppm (Risebrough, unpublished data). Migratory fish (tuna) carry up to 2 ppm of these same compounds in their gonads, and other marine mammals up to 800 ppm in their fat (Butler, 1969). In the coastal environment DDT and its residues range from undetectable levels to 0 to 5.4 ppm in oysters (Butler, 1969). Concentrations between these limits are highly variable locally, even within the same estuary.

From these few observations, it is possible to deduce that DDT and its residues are most probably distributed throughout the marine biosphere.

In spite of the paucity of useful data, some assumptions can be made for the marine environment, excluding estuaries, by assigning probable values for DDT residues to the biota of the open ocean and extending the calculations to a global basis. Such computations are valuable in that they identify potential sinks and provide order-of-magnitude estimates of most probable distributions. The following assumptions are made:

1.   The standing crop of plankton (plant and animal) is 3×10^9 metric tons (Menzel, unpublished data).

2.   The standing crop of fish is 6×10^8 metric tons, equal to 10 times the present annual fish harvest (Ryther, 1969).

3.   The concentration of DDT residues in plankton averages 0.01 ppm.

4.   The concentration of DDT residues in fish averages 1.0 ppm.

5.   A homogeneous distribution of DDT in the mixed layer has resulted from atmospheric transport from the continents.

By assigning the values in assumptions 3 and 4 to the calculated standing crop of organisms, 1 and 2, it is estimated that the total plankton now contain 3 metric tons and fish 600 metric tons of DDT residues, both insignificant fractions of the total annual input of these residues to the environment (10^5 metric tons). We proposed these estimates as upper limits to the size of the pool.

With the saturation level of DDT * in water of 1 ppb and the volume of the mixed layer (upper 100 m of the ocean) as 0.025×10^{24} ml, the surface waters of the ocean are capable of accommodating a load of 7.5×10^7 metric tons of DDT, or approximately 10 times the total production to date. There is no indication, however, that DDT introduced into the marine environment is uniformly distributed in the mixed layer. Enrichment is likely to occur in the sea's surface film, which contains fatty acids and alcohols. Predictions indicate (1) that DDT may be stripped from the film by bacteria and/or phytoplankton and thus enter the food chain; (2) that it adsorbs to airborne particles that sink through the water column (in this case the compound is probably ingested by grazing organisms which do not discriminate between living and inert particles); or (3) that it codistills with water, or is injected back into the atmosphere as aerosols and is

* Saturation levels of DDD and DDE have not been measured as yet.

redistributed, leading neither to a net increase or decrease in concentration at the surface.

Lacking any data for concentrations in the waters of the open sea, it is impossible to estimate directly how much is present that is not incorporated into living organisms. However, estimates of fallout in rain suggest that one-quarter of the world's production of DDT may have entered the ocean. Its areal distribution is probably uneven, dependent upon weather patterns and proximity to major sources of input.

If only 0.01 percent (100 metric tons) can be accounted for in pelagic marine organisms, 0.50×10^6 metric tons (one-quarter of total production) should be present in solution and in the bottom sediments. In order to balance input with accountable fractions, the surface mixed layer volume (0.025×10^{24} ml) should contain concentrations of approximately 5×10^{-12} g/ml, assuming a residence time of 5 years and an annual input of 0.25×10^5 metric tons of DDT/year.

Conclusions

The oceans are an ultimate accumulation site of DDT and its residues. As much as 25 percent of the DDT compounds produced to date may have been transferred to the sea. The amount in the marine biota is estimated to be in the order of less than 0.1 percent of total production and has already produced a demonstrable impact upon the marine environment.

Populations of fish-eating birds have experienced reproductive failures and population declines, and with continued accumulation of DDT and its residues in the marine ecosystem additional species will be threatened.

The decline in productivity of marine food fish and the accumulation of levels of DDT in their tissues which are unacceptable to man can only be accelerated by DDT's continued release to the environment.

Certain risks in the utilization of DDT are especially difficult to quantify, but they require most serious consideration. The rate at which it degrades to harmless products in the marine system is unknown. For some of its degradation products, half-lives are certainly of the order of years, perhaps even of decades. If most of the remaining DDT residues are presently in reservoirs which will in time transfer their contents to the sea, we may expect, quite independent of future manufacturing practices, an increased level of these substances in marine organisms. And if, in fact, these compounds degrade with half-lives of decades, there may be no opportunity to redress the consequences.

The more the problems are studied, the more unexpected effects are identified. In view of the findings of the past decade, our prediction of the hazards may be vastly underestimated.

Recommendations

We recommend a drastic reduction in the use of DDT substance as soon as possible. The recommendation of a recent Department of Health, Education, and Welfare report that the use of DDT (and DDD) be phased out by 1972 is sound, as indeed is the recognition given there to the necessity to continue use for the prevention of certain human diseases until an acceptable alternative is discovered (*Report of the Secretary's Commission,* 1969).

Some research has been devoted to such alternatives, and for many uses they exist. More research is needed, for which funds must be forthcoming. But research and the development of new, less-persistent pesticides is an inadequate solution or rather no solution at all. The pesticides may be less persistent, but they will still create more pests.* Other methods must be found. More research may profitably be devoted to biological control, chemical attractants and sterilants, and, in general, to the management of habitats where pests are a problem, either for agriculture or human health.

Consequently, we recommend that as the production and use of DDT are phased out, subsidies be furnished the developing countries to enable them to afford those alternatives that now exist (but are more expensive). Resources must then be made available, both at home and abroad, for an intensive, and cooperative research effort which will seek to develop not only less-persistent pesticides but also, it is hoped, an approach to the control of pests which is, as well, amenable to the stability of ecosystems.

* See discussion in section 2.5.3 in this report.

2.3.2

Mercury

For a discussion of the sources and production rates of mercury, see the report of Work Group 5. Mercury is one of the most serious environmental contaminants among the heavy toxic metals:

Elemental mercury and most compounds of mercury are protoplasmic poisons and therefore may be lethal to all forms of living matter. In general, the organic mercury compounds are more toxic than mercury vapor or the inorganic compounds. Even small amounts of mercury vapor or many mercury compounds can produce mercury intoxication when inhaled by man.* Acute mercury poisoning, which can be fatal or cause permanent damage to the nervous system, has resulted from inhalation of 1,200 to 8,500 $\mu g/m^3$ (micrograms per cubic meter) of mercury. The more common chronic poisoning (mercurialism) which also affects the nervous system is an insidious form in which the patient may exhibit no well-defined symptoms for months or sometimes years after exposure. (Stockinger, 1963)

Mercury is also dangerous when ingested in food. In Japan, 111 cases of mercury poisoning occurred (with 41 deaths) as a result of eating fish or shellfish taken from Minimata Bay. The bay had been contaminated by waste from an acetaldehyde production plant that had used mercuric oxide. Another outbreak occurred in Niigata City with 26 cases (and 5 deaths) due to the same cause (Irukayama, 1967).

Mercury's toxicity is permanent. In addition, when fish, shellfish, birds, or mammals containing mercury are eaten by other animals, the mercury may be absorbed and accumulated. Vertebrate animals reported as poisoned or killed by mercury, either under natural conditions or in experiments, include various species of fish, chickens, wood pigeons, Japanese storks, pheasant, quail, sheep, cattle, guinea pigs, dogs, rats, pigs, horses, cats, rabbits, and man (Carnaghan and Blaxland, 1957; Grolleau and Giban, 1966; Muto and Suzuki, 1967; Lofroth and Duffy, 1969). The maximum permissible level set for human food, including fish, is 0.5 ppm (based on the World Health Organization recommendation).

Industrial wastes and agricultural pesticides have caused severe mercury contamination in waters in Japan, Sweden, and

---

* Because of its high vapor pressure at room temperature, exposed mercury constantly emits vapors into the air.

the United States. While mercury use is increasing, there has been a general decrease in the use of mercury-containing pesticides in recent years. The Federal Water Quality Administration (FWQA) states that fish and water in 20 states in the United States are contaminated with mercury (*Congressional Record,* July 30, 1970). At present no mercury residue is allowed in foodstuffs in the United States. Fish containing unacceptable levels of mercury are confiscated or are not allowed to be sold in the United States. This metal is becoming of great significance as a persistent toxic environmental pollutant.

Since mercury is (1) highly toxic as a vapor or in its metallic or compound forms to most forms of life, (2) a nearly permanent poison once introduced into the environment, (3) biologically accumulated in organisms including man, and (4) being utilized in greater quantities throughout the world, it should be considered as a material that threatens to become critical in the world environment. Moreover, it should be reiterated that mercury is but one of approximately two dozen metals that are highly toxic to plants and animals. The emphasis necessitated by the case-study approach should not be misconstrued; indeed, the following recommendations may be appropriate for most or all of the toxic heavy metals, which also include lead, arsenic, cadmium, chromium, and nickel.

Recommendations

1. We recommend that all pesticidal and biocidal uses of mercury be drastically curtailed, particularly where safer, less-persistent substitutes can be used.
2. We recommend that data be obtained on the concentrations of mercury in selected organisms and on its effect on ecosystems.
3. We recommend that all industrial wastes and emissions of mercury be controlled and recovered to the greatest extent possible.
4. We recommend systematic and careful monitoring of uses of mercury and waste products containing mercury.
5. We recommend that world production and consumption figures for mercury be obtained.

2.3.3

Ocean Pollution by Petroleum Hydrocarbons
Sources of Petroleum Hydrocarbons in the Sea

At the present time the most conspicuously detrimental effects of oil pollution of the ocean are localized and are caused by accidental spills in near-shore areas. Those loci of concern potentially include the coastal zones of every continent and every inhabited island, so that the problem of accidental spills is of worldwide significance.

Although such accidents cause the most evident damage to ocean resources, they make up less than 10 percent of the estimated 2.1 million metric tons of oil that man introduces directly into the world's waters. At least 90 percent originates in the normal operations of oil-carrying tankers, other ships, refineries, petrochemical plants, and submarine oil wells; from disposal of spent lubricants and other industrial and automotive oils; and by fallout of airborne hydrocarbons emitted by motor vehicles and industry (see Table 5.10 in the report of Work Group 5). This oil in the oceans is now about 0.1 percent of world crude oil production, which reached 1.82 billion metric tons in 1969 (see Table 5.9 in the report of Work Group 5). If the possible fallout of airborne hydrocarbons on the sea surface is added, the total amount of oil and oil products contaminating the ocean at the present time may be as much as 0.5 percent of total production.

The magnitude of oceanic oil pollution is likely to increase linearly with the worldwide growth of petroleum production, transportation, and consumption. Since world crude oil production of 4 billion tons per year is predicted for 1980 (see Table 5.9 in the report of Work Group 5), the amount of oil entering the ocean directly may increase to 4 million tons by 1980.

The extent and character of the damage to the living resources of the sea from this "base load" of oil pollution is little known or understood. In the long run it could be more serious, because more widespread, than the localized damage from accidental spills.

Accidental Spills

All large accidental spills to date have occurred fairly near shore, and the spreading sheet of oil has drifted or has been blown by winds onto beaches and into shallow water areas. Efforts to contain and to dispose of the oil before it does extensive damage have been singularly ineffective. Various dispersing agents have been developed that break up the oil into small drops and cause it to sink below the surface. Some of these compounds are poisonous to ocean organisms, but even with a nontoxic dispersant the dispersed oil is much more toxic to marine life than is an oil slick on the surface. Even with our vast inventory of chemical agents, apparently the best and safest means of disposal is still absorption on chopped straw.

The danger of large-scale accidents increases with the size of tankers. Four 327,000-ton ships are already in operation; vessels of 500,000-dead-weight tons will soon be constructed, and 800,000-ton vessels have been projected within the next few years (Cooke, 1969). A single spill from one of the new large tankers could add 20 percent to the amount of petroleum entering the oceans in a single year.

The "Base Load" of Oil Pollution

It is likely that most of the oil entering the sea from ships, rivers, and the sea floor ends up in a narrow zone near shore at most only a few kilometers in width. Some of this oil will settle to the bottom. Certain fractions of the bottom-deposited oil will continue to disperse into shallow overlying waters for months or years. The highest oil concentrations will be near ports and harbors and in semienclosed seas such as the Mediterranean, the Black and North Seas, the Persian Gulf, and the Gulf of Mexico.

Submarine reservoirs of petroleum are likely to be found on the continental shelves of almost every continent, and the incidence of local contamination from underwater drilling and production on the continental margins will ultimately be widespread.

On the high seas, winds and ocean currents will bring about a convergence and retention of concentrations of hydrocarbons in the subarctic and equatorial convergence zones such as the Sargasso Sea. Workers from the Woods Hole Oceanographic Institution have found that oil globules and tar balls are more abundant in the Sargasso Sea than the Sargassum weed for which the sea is named (Blumer, 1969a).

Possible Fallout of Airborne Petroleum Hydrocarbons

The estimated emission of hydrocarbons of petroleum origin to the air each year is about 90 million tons (Goldberg, 1970; National Air Pollution Control Administration [NAPCA], 1970), roughly 40 times the amount of these substances entering the ocean directly from ships, shore installations, rivers, and the sea floor. It is known that a fraction exists as very small particles or becomes caught in rain. Much of this fraction may settle out on the surface of the ocean. If 10 percent of the hydrocarbons emitted to the atmosphere eventually find their way to the sea surface, the total hydrocarbon contamination of the ocean would be about five times the direct influx from ships and land sources. This quantity is expected to increase about as rapidly as the total petroleum production, that is, it should double by 1980.

Possible Biological Consequences of Oil Pollution

Depending upon their location, character, and concentration, petroleum hydrocarbon pollutants in the ocean can produce the following unwanted consequences:

1.   Poisoning of marine life filter feeders such as clams, oysters, scallops and mussels; other invertebrates; fish, and marine birds.
2.   Disruption of the ecosystem so as to induce long-term devastation of marine life from mass destruction of juvenile forms and of the food sources of higher species.
3.   Degradation of the environment for human use by reducing economic, recreational, and aesthetic values on both short and long-term bases.

Crude oil and oil fractions poison marine organisms through different effects:

1.   Direct kill through coating and asphyxiation or contact poisoning.
2.   Direct kill through exposure to the dissolved or colloidal toxic components of oil at some distance in space and time from the source.
3.   Incorporation of sublethal amounts of oil and oil products into organisms resulting in reduced resistance to infection and other stresses (one of the principal causes of death of birds surviving the immediate exposure to oil).

Observed Effects of Accidental Oil Spills

Oil from the *Torrey Canyon* tanker and the Santa Barbara oil well disasters killed many marine birds, but most of the damage to other animals and plants living in the intertidal zone and in shallow water is said to have been caused by the highly toxic detergents used to disperse the oil in the case of the *Torrey Canyon* (Holme, 1969) and by a layer of encrusting oil, which was often one or two centimeters thick, in Santa Barbara (Holmes, 1969).

Both these accidents occurred offshore in relatively deep water, and most of the toxic fractions of the oil may have evaporated or been dispersed to harmlessly low levels in large volumes of water before the oil drifted into the shallow waters near shore.

In contrast to these deepwater spills, the effects of relatively small oil spills in shallow water have been carefully observed, and here extensive damage was demonstrated. An accidental release of 240 to 280 tons of No. 2 fuel oil from a wrecked barge off West Falmouth, Massachusetts, in 1969 caused an immediate massive kill of organisms of all kinds—lobsters, fish, marine worms, and mollusks (Blumer, 1969b; Hampson and Sanders, 1969; Blumer, Souze and Sass, 1970, unpublished).

Effects of the Base Load of Oil Pollution

We have only fragmentary information about the biological effects of the base load of oil pollution in estuaries and coastal waters or on the high seas. One of the difficulties in assessing oil damage in coastal waters is that many other pollutants are also present in this zone, and it is hard to separate their different effects. Indeed, the effects may not be separable, but instead additive or mutually reinforcing.

The threshold of damage by direct poisonous effects is not now quantitatively known. Limited data indicate that filter-feeding mollusks (clams, oysters, scallops, and mussels) are killed by a mixture of oil and a relatively nontoxic dispersant in concentrations of 25 parts of oil per million parts of water. There are indications that if the oil were mechanically rather than chemically dispersed, similar concentrations of hydrocarbons would be toxic to filter-feeding mollusks.

One possibly serious effect of oil dispersed over wide ocean areas could arise from the fact that chlorinated hydrocarbons such as DDT and dieldrin are highly soluble in oil films. Measurements of the effects of a natural slick in Biscayne Bay, Florida, showed that the concentration of a single chlorinated hydrocarbon (dieldrin) in the top one millimeter of water containing the slick was more than 10 thousand times higher than in the underlying water.

About half of the pesticide in the water column was dissolved in the slick (Seba and Cochrane, 1969). We know that the small larval stages of fishes and both the plant and animal plankton in the food chain tend to spend part of the night hours quite near the surface, and it is highly probable that they will extract, and concentrate still further, the chlorinated hydrocarbons present in the surface layer. This could have seriously detrimental effects on these organisms and their predators.

Recommendations

1. We recommend that the following research be undertaken:

    a. Determination of the levels of concentration of different petroleum fractions in dissolved or colloidal form which are toxic to sensitive marine organisms such as filter-feeding clams, oysters, and scallops, and to the larval stages of other invertebrates and fishes. The physiological mechanisms of damage also require further study.

    b. Elucidation of the possible effects of different petroleum fractions on the communication and information-gathering systems and on the reproductive, feeding, and defensive behavior of marine invertebrates and fishes.

    c. Determination of the extent of concentration of persistent pesticides and other chlorinated hydrocarbons in oil films, slicks, globules, and masses on the surface in the open ocean, and the effects on birds, fish and plankton.

2. We recommend that more information be compiled on the distributions of petroleum and other hydrocarbons at the surface of the open sea, in the water column, and the bottom sediments of estuaries and near-shore areas. This will require not only monitoring of these distributions but also studies of the rates and mechanisms of evaporation, solution, dispersion, and bacterial or physical oxidation of different petroleum fractions, and the rates of fallout of hydrocarbons in particulate form from the atmosphere onto the sea's surface.

3. We recommend that several lines of technical development designed to reduce or remove direct oil pollution in the sea be pursued. These include

    a. Better methods of removing oil from large accidental spills. The use of even nontoxic dispersants is undesirable, because the oil is broken into fine droplets which poison the filter-feeding organisms that may ingest them. Burning large quantities of oil at the sea surface near the shore may produce severe local air pollution. Containment of the oil where it can do least harm, followed by physical removal, is most satisfactory.

b. More effective and economical antipollution measures for refinery and petrochemical operations.

c. More economical and easily used techniques, which do not involve the discharge of oil residues in the sea, for cleaning the bilges and fuel tanks of merchant ships.

4.   We recommend that regulations and other institutional changes be instituted to reduce or eliminate the discharge or leakage of waste oils and greases into rivers, lakes, estuaries, and coastal areas.

5.   We recommend that more effective international control measures for oil-carrying tankers be developed

a. To ensure that all tankers use "load on top" or other antipollution procedures.

b. To prevent strandings and collisions. This may require monitoring and control of tanker tracks at all times, just as transoceanic aircraft are now regulated.

6.   We recommend that international standards for safe procedures and an international mechanism for their enforcement be developed because many future offshore oil drilling and production operations can be expected off the coast of countries that are not technically equipped to enforce effective antipollution regulations.

2.3.4

Phosphorus

Introduction

Of the major nutrients—nitrogen (N), phosphorus (P), and potassium (K)—which man introduces into the environment, phosphorus can cause the most serious pollution problems. Eutrophication of lakes is already a major local problem; eutrophication of estuaries is a potential global problem.

Some phosphorus production data are given in the report of Work Group 5. Verduin has found that most phosphorus enrichment of surface waters comes from sewage treatment plants and only secondarily from agricultural fertilizers (Verduin, 1966). Table 2.8 shows the relative amounts of phosphorus entering rivers from municipal raw sewage, and urban and rural runoff. Primary sewage treatment removes about 10 percent of the phosphate and secondary treatment about 30 percent (American Chemical Society [ACS], 1969). Although much of the phosphate may settle during some phase of treatment, it is released back into the effluent during subsequent sludge digestion. In view of the small nutrient mass in the sludges, their separation from the main stream of the sewage makes little difference to the eventual nutrient enrichment of ocean areas.

Nutrient runoff from agricultural land is not yet a major global problem. Most fertilizers are applied in developed countries, and consequently the estuaries in these countries are the principal sinks for agricultural nutrients. Phosphorus from farmland enters streams primarily through runoff from feed lots and erosion of topsoil. Soil particles have a strong affinity for phosphate, thus reducing the amount of phosphorus in runoff.

Eutrophication of Lakes

The eutrophication of lakes is a serious problem in practically all countries and may serve as a model for what is likely to happen to estuaries and possibly the coastal ocean if present trends continue. Many lakes throughout the world are overnourished (eutrophic) or are becoming overfed by effluents in which the key ingredient is phosphorus from domestic sewage, eroded soil, and farm manure.

Lakes that have been clear and clean for thousands of years have become repulsive and odorous within ten years after man-made effluents are introduced (Hasler, 1969). Excess nutrients change the algal community from one of great diversity of species to one of a few species which cause nuisances. Costs of purifying water for human use rise, property values drop, and recreational features deteriorate. When eutrophication progresses to its extreme stages, the water becomes deficient in dissolved oxygen, and all useful species (including fish) die. Eutrophication in lakes can, however, be alleviated by stopping waste discharge (for example, the cases of Madison, Wisconsin, lakes, and Seattle's Lake Washington described in Hasler, 1969).

Although eutrophication is caused by a mix of nutrients,

Table 2.8
1968 U.S. Mean Phosphorus Content of Runoff

Source	Percentage of Total U.S. Discharge[a]	Magnitude (metric tons)[b]
Municipal raw sewage	60	262,000
Urban runoff	23	100,000
Rural runoff	17	74,000

[a] Municipal raw sewage and urban runoff percentages were computed using the 17 percent figure for rural runoff given in Federal Water Pollution Control Administration (FWPCA), 1968, and the following data from American Chemical Society (ACS), 1969: of the total U.S. urban draining, raw sewage constitutes 20,000 lb/sq mile/yr, and all other sources contribute 7,800 lb/sq mile/yr.

[b] These magnitudes were computed using the percentages in the table and the following data from ACS, 1969: in U.S. surface waters, there are 280 million lb/year of phosphates from household detergents and 680 million lb/year from other sources for a total of 960 million lb/year or approximately 436,000 metric tons.

potassium is usually present in excess and tends to remain in the water, while nitrogen levels can be supplemented through the biological fixation of atmospheric nitrogen. Phosphorus tends to be precipitated in sediments, and phosphorus levels cannot be supplemented. Thus, phosphorus is the most likely substance limiting algal growth, and it serves as the best general indicator of nutrient enrichment (Rohlich, 1969; ACS, 1969; Vollenweider, 1969).

One of the most graphic examples of lake eutrophication in the United States is the case of Lake Erie. The Federal Water Pollution Control Administration (FWPCA) estimated rural runoff phosphorus at 0.1 metric ton per square mile per year and urban runoff at 0.25 metric ton per square mile in the Lake Erie drainage basin (FWPCA, 1968). Municipal waste accounted for 70 percent on the phosphorus contributed to Lake Erie in 1967 and is projected to contribute 80 percent in 1990.

Estuaries and Coastal Ocean Areas

Estuaries are semienclosed basins where water flowing from continents mixes with surface and subsurface ocean waters. Water circulation in an estuary is a two-way process. Surface layers of less saline water flow generally seaward over subsurface layers of more saline water flowing generally landward. The subsurface flow replaces water and salt entrained in surface waters, while extensive mixing occurs across the pycnocline (the layer of strong density difference separating the surface and subsurface layers). Estuarine-type circulation is not restricted to estuaries but is typical of most coastal ocean areas where freshwater inputs from rivers and precipitation exceed water loss through evaporation.

This circulation pattern produces several important effects. It results in estuaries and coastal ocean areas accumulating river-borne sediments and wastes, as well as wastes dumped by adjacent urban areas and sediments (commonly sands) moving along the coast. (Harbor dredging and shipping facility construction improve an estuary's sediment-trapping efficiency and also present the problem of disposing of the dredged wastes.) More important, as a result of this circulation pattern, estuaries trap nutrients. This results in the well-known productivity of coastal waters. Surface organisms feed on the nutrients (especially nitrogen and phosphorus), die, sink to the bottom, and decompose. When productivity is excessively large, this may exhaust the dissolved oxygen in the bottom waters, killing all normal benthic organisms except bacteria. Landward-moving bottom waters return the released nutrients to the surface, supporting the growth of more phytoplankton. Thus, nutrients tend to be retained and recycled in the estuary (or coastal ocean water) rather than move seaward to mix with the open ocean.

Estuaries receive riverborne wastes from upriver and adjacent communities and industries as well as the natural discharge of nutrients and solids from river and rain. Table 2.9 illustrates the relative size of each contribution for a hypothetical community of 10 million people covering an area about 12,000 square miles, on a river discharging about 500 cubic meters per second (15,000 cubic feet per second). Each volume of river water mixes with 20 volumes of nutrient-carrying subsurface water. (Note: No allowance has been made for the industrial wastes commonly discharged directly into the estuary.) This table illustrates the local origin of most estuarine problems involving excessive nutrient discharge and accompanying ecological imbalance. The dominant source of phosphate is sewage (primarily from detergents containing phosphate), while nitrogen in various forms is supplied in

Table 2.9
Example of Possible Nutrient Discharges to Estuaries from Various Sources
(Metric tons per year)

	Nitrogen (N)	Phosphorus (P)
Sewage:[a] population of 10 million, per capita generation 400 liters/day	6,000	15,000
River water:[b] discharge 500 m³/sec (approx. 15,000 cfs)	300	7
Subsurface seawater:[c] 20 volumes of seawater mixing with each volume of river water	6,000	900
Storm-water runoff:[a] 75 cm (30 inches) per year area	6,000	800

Sources:
[a] Weibel, 1969.
[b] Bowen, 1966.
[c] Ketchum, 1969.

nearly equal volumes by sewage treatment plants, storm-water runoff, and subsurface seawater (Verduin, 1966).

Most of the world's large urban centers (in both developed and less-developed countries) are located on estuaries and discharge their wastes into the coastal waters (Cronin, 1967). Since estuaries are the permanent residence, passage zone, or nursery area for about 90 percent of commercially important fish, pollution here could have far-reaching effects (McHugh, 1967; FWPCA, 1970). For example, over 20 percent of the world's minimal protein needs could come from the sea by the year 2000, provided estuaries are still healthy (Ricker, 1969). Moreover, national pollution can have international effects. For example, the pollution of Alaskan and Canadian estuaries, the principal nursery of Pacific salmon, would affect not only the United States and Canada but also Japan and the USSR, who fish the salmon on the high seas.

The concentration of industrial nations around the North Atlantic basin and the common fate of estuaries in industrialized nations lead us to recommend careful investigation and consideration of international monitoring. Yet, because of the usual local origin of estuarine problems, effective control is most likely to be accomplished unilaterally.

Recommendations

1.   We recommend that nutrients in areas of high concentrations, such as sewage treatment plants and feedlots, be reclaimed and recycled.

2.   We recommend that the dumping of industrial wastes into sewage systems be restricted, so that toxic wastes do not interfere with nutrient recovery and recycling.

3.   We recommend that nutrients not be used in materials that are discharged in large quantities into water or air. We recommend, for example, that phosphates in detergents be replaced with new materials, being certain that the substitute does not itself create a new problem.

4.   We recommend that the institutional structures responsible for defining, monitoring, and maintaining water quality standards over large areas be improved. The multiplicity of authorities involved in river basins, estuaries, and coastal oceans makes effective control nearly impossible.

2.3.5

Wastes from Nuclear Energy

It has taken our full efforts to probe in some depth these few questions. We decided deliberately to omit consideration of others of great importance. One of these is the problem of perpetual management of the large quantities of radioactive wastes that are by-products of nuclear power.

No other environmental pollutant has been so carefully monitored and contained. Yet as we look back on the intense examination by Work Group 1 of the effects of the products of fossil fuel combustion, we have become aware of our neglect of a different class of pollutants that will grow greatly in quantity in the next 30 years.

We call to the attention of one of the supporters of SCEP, the AEC, our decision to omit this item from our agenda and our concern about the subject.

Recommendation

We recommend that an independent, intensive, multidisciplinary study be made of the trade-offs in national energy policy between fossil fuel and nuclear sources, with a special focus on problems of safe management of the radioactive by-products of nuclear energy, leading to recommendations concerning the content and scale and urgency of needed programs.

2.4
Information Needs
2.4.1
Introduction

The need for better information on sources, effects, and distributions of pollutants is documented in the preceding case studies. Indeed, it can be argued that the single greatest contributor to environmental pollution is ignorance. Furthermore, data sufficient to raise an alarm are not adequate for designing and operating a management system. It is in the nature of any study of the environment to seek effective means whereby the information needed now may be expeditiously obtained and at the same time to devise a system that will furnish such information at regular intervals in the future. This section presents a set of specific recommendations which, if implemented, would provide information needed to improve management decisions and research designs.

2.4.2
Information on Distribution

Of the two proposals that follow, the first suggests simplified methods for the collection of data pertinent to an analysis of ecosystem function, while the second reflects the great gaps in our knowledge of the present concentrations of potentially toxic pollutants in the marine environment. The design of more ambitious programs in these two areas of study is presented in the report of Work Group 3. Hence, the title of this subsection is intended to suggest not only a gap in our knowledge of the location of various pollutants in, and their impact on, the biosphere but also different approaches to gathering that information.

Ecological Systems

A considerable body of data exists on ecological systems. If more of it had been collected in an organized, systematic fashion, and if more of it were critically aimed at measuring ecosystem function, we would not now suffer so severe a handicap in environmental evaluation. These deficiencies can be remedied on a local scale without waiting for organized national and international programs. Moreover, inexpensive but critical data, systematically collected in a number of places, would become useful in regional and global evaluation.

Specific suggestions follow for the collection of data that are useful in the diagnosis of ecosystem function. An important characteristic of these methods is that data can be collected in different ways at different places using various species or species groups. The major constraints are (1) that the method chosen be used consistently in any one place, and (2) that measurements be made systematically over time. The following list is suggestive, not exhaustive, of the variety of measures to be considered, and it is organized according to the global ecological effects described earlier.

DETECTING DAMAGE TO PREDATORS. (1) In local and regional populations of land and marine birds, especially for species that have recognized juvenile forms, the reproduction and survival of the young to maturity (animal recruitment) need to be counted. (2) In commercial and game-fish populations that normally contain many year classes, catches should be sampled to estimate recruitment of the youngest-year classes. (3) Native herbivorous insects should be raised and set out at appropriate life stages to attract insect enemies. Species should be used that are not pests and are themselves so sparse that their enemies are virtually impossible to find. Samples of eggs can be observed for the frequency of egg parasites, and insect larvae can be set out to attract parasites of later growth stages. Moths of the silkworm family exist throughout the temperate and tropical zones but rarely become numerous. Aphids are another group that should be used intensively.

DETECTING LOSS OF SPECIES. (1) Plankton samples of algae can be sorted to species and counted, a task that would provide an estimate of diversity and species abundance. Zooplankton are easier to sort to species and are just as informative. Both should be done, but the minute plants and animals should be collected in nets with different mesh sizes. (2) Samples of soil are easily examined for diversity among the soil animals. With an overhead light as a heat source, the soil in a funnel, and water or alcohol in a beaker below, hundreds to thousands of animals crawl downward and fall into the beaker. (3) Inverted light traps collect hundreds of individuals and dozens of species of flying insects. Records over time give good approximations of the abundance and diversity of species that are attracted to light.

MISCELLANEOUS MEASURES. (1) Tree leaves, especially at the moment of leaf fall, can be examined for the relative area removed or damaged. The latter averages around 6 percent in a healthy forest ecosystem and may prove to be a very sensitive measure of the functional success of natural control of leaf-eating insects. (2) Oxygen depletion in deep water, especially in August, provides an estimate of the accumulated burden of organic matter that has sunk from upper levels during the summer. Useful where the water is stratified (warm above, cold below, or fresh above, salty below), low oxygen levels are a signal of overenrichment and danger to game fish. (3) The growth of trees (tree rings) can be followed annually as well as backward in time. Enough estimates to obtain a good local average will reveal trends in growth that are moderately pronounced. The latter becomes a measure of annual forest production.

Sampling the Marine Environment

Data are available about the emissions of various pollutants into the atmosphere, the rivers, the estuaries, and the ocean through rivers and coastal sewage outfalls. We have some information concerning the resulting concentrations on land and in freshwater, some about the resulting concentrations and effects of pollutants in estuaries and lagoons, but very little about the coastal waters. Open ocean concentrations of the various pollutant substances introduced by man have not yet been measured.

We can show that many pollutants are available to the oceans through various routes, such as the atmosphere, runoff, dumping, and ships. If we knew the form in which a particular substance reached the ocean (gas, solute, particle, liquid), it might be possible to calculate its concentration and distribution in various areas and depths of the ocean, provided that it did not undergo any biological or chemical transformation. However, it is precisely because these substances do take part in the chemical and biological cycling in the ocean that they are of serious concern. Because the manner of cycling of these substances in the open ocean is not yet understood, we cannot make accurate estimates of their distributions, concentrations, and effects.

Analysis of a small number of samples collected from the open ocean would give immediate clues as to the amounts of various substances the ocean has received, the paths through which they have entered, how long they persist, how they are transformed, how they are concentrated, and what their effects are now and are likely to be in the future.

It seems worthwhile to carry out a brief (1-year) and simple sampling program of the marine environment. Enough is known of the atmospheric circulation over the ocean, the water circulation, including convergences, divergences, and upwelling, about the areas of high and low primary productivity and biomass concentration, and about the major fish populations, so that a coarse array of samples (at most 1,000) would be useful in the definition of the important pollution problems in the oceans.

Of these samples (biomass, sediment, water, surface film, air), the most important will be those taken from the biomass (fish and plankton). For, since the biomass may concentrate many of these substances from the water, pollutants may be most easily detected there.

Many of these samples, especially fish and plankton, can be obtained from collections already at hand or from those being compiled by existing programs. Some samples, such as oil slicks, can be obtained through minor additions to present programs. Others will be more difficult and expensive.

The choice of materials to be assayed as potential threats to marine ecosystems may be sought on the basis of the following categories:

1. *The minerals or solid phases that result from the combustion of fossil fuels and industrial activities.* The production of energy through the burning of coal, oil, and natural gases is accompanied by the release of such gases as carbon dioxide, sulfur dioxide, and nitrogen oxides, as well as by the atmospheric entry of many of the constituents of the fuel itself (vanadium, nickel, and copper). A measure of fossil fuel combustion may be sought in concentrations of carbon (soot, cokey balls) in wind dust loads, rains, and lake and oceanic sediments. The elemental

carbon may act as a most important tracer of the paths of materials introduced to the atmosphere and the oceans as a result of energy production.

2. *Elements introduced to the atmosphere or rivers in amounts that are of the same order of magnitude as those brought to the world's rivers by natural weathering processes.* Cadmium, mercury, nickel, lead, and vanadium appear to be five heavy metals in this category.

3. *Synthetic organic chemicals, produced in large quantities, which can interfere with the metabolic processes of marine organisms.* Initially, one can consider such materials as the dry-cleaning solvents, perchlorethylene and trichlorethylene, freon, gasoline, and the chlorinated biphenyls.

2.4.3

Information on Effects

The protection and enhancement of our water resources require not only a continuing measure of what is being added to them but also a thorough knowledge of those concentrations of wastes and materials that are not harmful even with continuous exposure. The latter is essential for the definition and detection of pollution, for the determination of the objectives and goals of waste treatment and pollution abatement, and for the establishment of water quality standards.

Toxicity studies conducted over the past 40 years have not supplied data essential for the setting of definitive water quality standards. Little is known of the requirements of the most sensitive species, thus the results do not indicate safe levels for the biota. To meet this problem, well-coordinated, thoroughly planned, and adequately funded national and international programs are essential.

The toxicity of acute doses of many pollutants is known for humans and, in some cases, for lower organisms in both terrestrial and aquatic ecosystems. Although the long-range effects of continued exposure to some toxic materials are also known for man, these data are almost nonexistent for other organisms.

Plants and animals that inhabit the waters of the world are in intimate contact with their environment. There is a phase boundary between organism and water across which toxic substances (especially the soluble and colloidal forms) transfer efficiently and directly. Because aquatic organisms may easily accumulate added pollutants throughout their lifetimes, the long-range effects of continued additions of these materials must be determined soon if man desires to maintain or to improve the life-supporting and aesthetic qualities of the world in which he lives.

The development of information essential for the setting of water-quality standards for the protection of aquatic life resources requires the determination of (1) toxicants that may be added in significant quantities to the hydrosphere; (2) the important species of the aquatic biosphere, based on their economic or recreational values, their importance in the maintenance of food chains, ecosystem balance, or in the production of oxygen or food for man; and (3) maximum concentrations of wastes or materials that are not harmful with continuous exposure.

The selection of test organisms must be given careful consideration, because there is great variation in the toxicity of the same materials to different species and life stages and of different materials to the same species. In many aquatic organisms the early life stages, including the egg stage, are particularly sensitive to toxic materials or to other environmental stresses, and representatives of this group should be included in a test program. Other organisms, especially benthic forms and planktonic filter feeders, show marked propensities for accumulating precipitated and coprecipitated materials and thus may accumulate sufficient toxic particles to cause their own deaths or to pass concentrated amounts of the pollutants along the food chain to higher trophic levels. Representatives of this group should be tested.

Some adult organisms with high metabolic rates are especially susceptible to stresses from increased pressure and salinity or from decreased pH and oxygen levels. Trout and tunas are representatives of this group. The phytoplankton are important as primary producers (and concentrators) and in helping to maintain the oxygen-carbon dioxide balance in water. These plants should receive high priority in the test program. Finally, some pollutants, including DDT and other fat-soluble materials, are concentrated with increased trophic levels. Sublethal toxicity studies of DDT, aldrin, toxaphene, and petrochemical components should thus be conducted with carnivorous fish, sharks, and porpoises.

The results of laboratory tests must be corroborated in the field to determine their adequacy under natural conditions, where the additional stresses of parasites, disease, predators, competitors, and other environmental factors are operative. Studies must also be made of the effects of concentrations of wastes in bottom sediment and of the relation between the concentration of a material in the water, in bottom sediment, in the bodies of organisms, and in passage through the food chain.

Even with the use of shortcuts by which long-term studies are carried out through two generations with only the most sensitive species, considerable time will elapse before needed data are compiled for all important wastes. In the meantime, all available data should be used to set tentative standards.

From the international viewpoint, the persistent pollutants are of prime importance. When water quality requirements are determined for these materials, they will in general be applicable worldwide, making it possible to set international water quality standards for open ocean areas. Because of differences in areal biota, standards will vary for estuaries and coastal waters; however, it is believed that there will be sufficient similarity to render beneficial the exchange of data on an international basis.

2.4.4

Information on Sources

Man now produces more than a million different kinds of products, both as waste and as final products that eventually end up as waste. Certain of these are easily identified as threatening to the environment; for some, production and use figures are given in this report. The problem is how to maintain an adequate surveillance of all major pollutants. While a complete documentation of the required activities is not feasible, the following steps will greatly improve our ability to recognize new problems sooner.

Surveys of New or Rapidly Expanding Activities

Especially in industry, innovation and change should come under the scrutiny of a data-gathering-and-evaluating group.

Tentative Identification of Potential Threats

All new products and activities similar to existing products and activities that are known to be deleterious to the environment should be noted for special study. For example, these products would include all of the chlorinated hydrocarbons. Major land changes and plans to build new industries, especially along water, should also be included. The total number of categories to be assessed will be large but very much smaller than the totality of change.

Immediate Research on Environmental Effects

For new wastes or products that will enter the environment, research should focus on (1) toxicity or other biological action at low and chronic levels, (2) persistence in the environment, and (3) the time response of the potential target.

Priority for Intensive Study

Depending upon the results of this research, upon the estimated time for recovery should damage occur, and upon current and projected rates of release, a high priority for intensive study can be given. Additionally or alternatively, the burden of proving the material safe could be placed on the producer, with production prohibited until such proof is satisfactory.

2.5

General Considerations of Ecology and Life-Style

In addition to acting immediately on many specific items, such as those recommended in the preceding section, man must soon solve some very basic problems that adversely affect many of his activities. A few are described here which require solutions that will demand far-reaching changes in some human values and activities.

2.5.1

Population Growth

For some time, and indefinitely into the future, additional numbers of people can be accommodated only by a more intensive use of land and consumption of a larger share of the earth's resources. This prognosis implies less room for other forms of life. Sooner or later, we must decide how much life to displace. Population will increase whether we decide to eliminate all other natural life or whether we try to preserve a biosphere that still provides us with most of our major needs. At the current rate of population increase such a biosphere may not last for more than 100 years.

2.5.2
Ecological Growth

An indefinitely rising material standard of living has nearly the same effect on the biosphere as an indefinitely rising population. As with population, this problem must also be solved sooner or later. A shift involving increased emphasis upon spiritual, aesthetic, and intellectual components in a standard of living remains the only way that standards could improve indefinitely. At present, the Gross Domestic Product (an approximation for the Gross National Product) for nations and its homologue for individuals stand as the most generally used criteria of status. A basic change in values in which increase in material wealth is not so highly rated must accompany any solution to the problem of ecological growth.

2.5.3
Pesticide Addiction

Realization that the use of pesticides increases the need to continue their use is not new, nor is the awareness that the constant use of pesticides creates new pests. For many of our crops on which pesticide use is heavy, the number of pests requiring control increases through time. In a very real sense, new herbivorous insects find shelter among our crops where their predator enemies cannot survive.

Fifty years ago most insect pests were exotic species, accidently imported to a country lacking their natural enemies. More recently many of the pests, including especially the mites, leaf-rolling insects, and a variety of aphids and scale insects, have been indigenous.

Thus pesticides not only create the demand for future use (addiction), they also create the demand to use more pesticide more often (habituation). Our agricultural system is already heavily locked into this process, and it is now spreading to the developing countries. It is also spreading into forest management. Pesticides are becoming increasingly "necessary" in more and more places.

Before the entire biosphere is "hooked" on pesticides, an alternative means of coping with pests should be developed. At the very least this will include crop breeding for resistance, changes in crop patterns that may not be so well suited to mechanization, massive developments in the techniques of managing predators, parasites, and diseases for use in control, relaxed acceptance values for blemished fruit, and a general acceptance of low levels of insects as a situation we do not have to fight.

2.5.4

Communication and Education

As our technology becomes increasingly powerful, it is increasingly urgent that we educate our children (and ourselves) to the dangers of misusing the environment. This can be done both in the schools as a part of the regular curriculum and in special programs designed to familiarize the student with the functions of natural and man-altered ecosystems. For the latter, "ecological classrooms" can be set aside to be used by the public to give meaning to its growing awareness of the environment. Similar areas are needed by scientists so that base lines can be established against which to measure change.

The present surge of activity in the classrooms, in the mass media, and in various park systems attests to the general recognition of these needs. In many cases, however, and especially in the schools, a critical lack of teaching material, experience, and expertise has forced many individuals to improvise programs whose quality and soundness are less than satisfactory. A much greater effort is needed to bring professional competence into the development of these programs.

No change in life-style will occur unless the need for change is broadly recognized and accepted. It remains the responsibility of those who study environmental problems and understand their dimensions to convey this need to the public. It is also their responsibility to determine that information organized in a form that can be understood by the public accurately reflects the evidence and scientific arguments taken from their professions. Several existing organizations, appropriately expanded, can aid greatly in this process of bringing information to the public and to decision makers. Imagination and energy are required for the

development of additional means for increasing enlightened
public discussion of critical environmental problems.

2.6

Appendix: Carbon Cycle in the Biosphere*

About half of the carbon dioxide released into the atmosphere
by the burning of fossil fuel has disappeared. The two large
carbon "pools" that interact with the pool in the atmosphere
are the vegetation of land surfaces and the carbonate-bicarbonate
system in the oceans (together with living matter in the oceans).

Slightly differing estimates from several sources (FAO Year-
books, 1951–1969; Ryther, 1969; Olson, 1970; Whittaker and
Likens, unpublished) were combined into a single set producing
the summary given on Table 2.A.1. Although the accuracy of
these estimates is poor, they are the best that are available.

For the purpose of analyzing the dynamics of interaction
between the vegetation and the atmosphere, the organic carbon
pools were divided into "short-lived" and "long-lived" components. The former includes leaves, litter, short-lived animals,
and most algae, while the latter includes wood, large roots, upper
soil humus, and dead organic matter in the oceans (see Table
2.A.2). While these estimates are speculative, they represent an
approach to data analysis that usefully interrelates ecology and
meteorology. Better estimates must await new research and model
building that may be generated from this analysis.

The results suggest that the organic carbon pool, with a
mean residence time of the order of 60 years (terrestrial wood,
humus, and so forth), is about twice the size of the atmospheric
pool and could easily have absorbed a significant portion of the
carbon dioxide that has left the atmosphere. Nevertheless, the

---

* This appendix presents the findings and recommendations of a SCEP group
whose members were J. B. Hilmon, C. D. Keeling, Lester Machta, Jerry S.
Olson, Roger Revelle, Walter O. Spofford, and Fred Smith. Jon Machta of
the University of Michigan assisted this group in helping to define the components to be put into a computer model.

The data presented here are not scientifically firm; numbers frequently
are based on educated guesses. Nevertheless, this presentation of these numbers
is valuable in demonstrating an interdisciplinary analytical technique and is
included as an illustrative approach to the problem of measuring the various
carbon pools.

Table 2.A.1
Organic Carbon and Its Rates of Production[a]
(Living and dead, excluding incipient fossil deposits)

	Area[b] 10^6 km²	Organic Carbon Pool[b] 10^9 metric tons	Production/Year[b] 10^9 metric tons
Reservoir			
Forest and woodland	48	1,012	36
Grassland and tundra	38	314	9
Desert and semidesert	32	59	3
Wetlands	2	30	2
Glaciers and barren	15	0	0
Agricultural	15	165	6
Total terrestrial	150	1,580	56
Oceanic	361	703	22
Burning fossil fuel (1970)			4
Atmospheric pool		683	

[a] Resistant humus and other material with decay rates of 0.001 per year or less have been omitted. Production is "net primary production," that is, production from photosynthesis minus plant respiration; it represents the yield to animals and decomposers.

[b] The numbers shown here are intermediate values from the several sources listed in the text. Although these sources present similar estimates, their combined accuracy is not regarded as high; procedures for obtaining global estimates for characteristics of vegetation are still primitive.

residence time is short enough so that decomposition would soon be returning larger amounts to the atmosphere.

Recommendations

A number of recommendations emerged from this effort, derived to a considerable extent from the fresh views we obtained through interdisciplinary discussion.

1. We recommend that an organized information system, such as that being started under the International Biological Program, be developed and maintained to eliminate uncertainties in comparing techniques and results from scattered sources.

2. We recommend that differences in experience among teams of experts of varying backgrounds be used to test whether their

Table 2.A.2
**Subdivision of Organic Carbon into Materials
with Short or Long Residence Times**

Reservoir	Organic Carbon Pool		Mean Residence Time (years)	
	Short	Long	Short	Long
Forest and woodland	62	950	3.1	59
Grassland and tundra	11	303	2.6	63
Desert and semidesert	4	55	2.2	39
Wetlands	1	29	1	36
Agricultural	4	161	1	80
Total terrestrial	82	1,498	2.3	62
Oceanic	2	701	0.1	701

Mean residence of carbon in the atmosphere: 8.8 years.
Note: The numbers shown here are not data but intelligent guesses of the way the estimates from Table 2.A.1 can be subdivided. Definitions of "short" and "long" residence times are given in the text.

calculations can be checked out and improved. Japanese, Scandinavian, Australian, Central European, and Inter-American groups are already developing the necessary team approaches.

3.   We recommend that radiocarbon and other techniques for determining age, together with studies of decomposition rates, be developed for the various components of dead organic matter. Total soil profile determinations of organic matter are exceedingly scarce. In addition, the total amount of living and dead material is not known, and its residence time is also largely unknown.

4.   We recommend that more detailed analysis be given to second-stage modeling. While a two-component (short versus long-lived) approach permits the development of initial models of the interaction between the atmosphere and vegetation, both fractions are parts of a broad spectrum of materials of various longevities.

5.   We recommend that the effects of deforestation be given serious study. The conversion of forests to grassland and cropland and the harvesting of forests both greatly reduce the pool

of organic carbon in the terrestrial system. Since the rise of agriculture, man has removed an amount of carbon that appears to be about half the size of the atmospheric pool and added it to the air as carbon dioxide. A historic analysis of periods of intense deforestation and of climatic variations may give empirical insight into the effect of carbon dioxide on the climate. Similarly, removal of the great tropical forests could have an impact on the dynamics of the carbon cycle at least as serious as the burning of fossil fuels.

References

Acree, F., Berova, M., and Bowman, M. C., 1963. Codistillation of DDT with water, *Journal of Agriculture and Food Chemistry, 11*.

American Chemical Society (ACS), 1969. *Cleaning our Environment: The Chemical Basis for Action* (Washington, D. C.: ACS).

Anderson, D. W., Hickey, J. J., Risebrough, R. W., Hughes, D. F., and Christensen, R. E., 1969. Significance of chlorinated hydrocarbon residues to breeding pelicans and cormorants, *Canadian Field Naturalist, 83*.

Bailey, T. E., and Hannum, J. R., 1967. Distribution of pesticides in California, *Proceedings of the Sanitary Engineering Division of the American Society of Civil Engineers, 93*.

Bitman, J., Cecil, H. C., and Fries, G. F., 1970. DDT–Induced inhibition of avian shell gland carbonic anhydrase: a mechanism for thin eggshells, *Science, 168*.

Blumer, M., 1969a. Oil pollution of the ocean, *Oil on the Sea*, edited by D. P. Hoult (New York: Plenum Press).

Blumer, M., 1969b. Oil pollution of the ocean, *Oceanus, 15*.

Blumer, M., Souze, M. G., and Sass, J., 1970. Hydrocarbon pollution of edible shellfish by an oil spill, Woods Hole Oceanographic Institute reference 70–71 (unpublished).

Bowen, H. J. M., 1966. *Trace Elements in Biochemistry* (London and New York: Academic Press).

Bowman, M. C., Acree, F., and Corbett, J., 1960. Insecticide solubility, solubility of carbon-14 DDT in water, *Journal of Agriculture and Food Chemistry, 8*.

Butler, P. A., 1964. Commercial fishing investigations in effects of pesticides on fish and wildlife, *Fish and Wildlife Service Circular, 226*.

Butler, P. A., 1966. Pesticides in the marine environment, *Journal of Applied Ecology, 3* (Supplement).

Butler, P. A., 1967. Pesticide residues in estuarine mollusks, *National Symposium on Estuarine Pollution* (Stanford University).

Butler, P. A., 1969. Monitoring pesticide pollution, *Bioscience, 19*.

Carnaghan, R. B. A., and Blaxland, J. D., 1957. Toxic effect of certain seed dressings on wild and game birds, *Veterinary Record, 69*.

Chemical Economics Handbook, 1969. (Menlo Park, California: Stanford Research Institute.)

164 Work Group 2

Congressional Record, July 30, 1970 (Washington, D.C.: U.S. Government Printing Office).

Conney, A. H., 1967. Pharmacological implications of microsomal enzyme induction, *Pharmacological Review, 19.*

Cooke, R. F., 1969. Oil transportation by sea, *Oil on the Sea,* edited by D. P. Hoult (New York: Plenum Press).

Cronin, L. E., 1967. The role of man in estuarine processes, *Estuaries,* edited by G. H. Lauff (Washington, D.C.: American Association for the Advancement of Science).

Davis, K. P., 1954. *American Forest Management* (New York: McGraw-Hill).

Federal Water Pollution Control Administration (FWPCA), 1968. *Cost of Clean Water,* Vol. III (Washington, D.C.: U.S. Government Printing Office).

Federal Water Pollution Control Administration (FWPCA), 1970. *National Estuarine Inventory* (Washington, D.C.: U.S. Government Printing Office).

Food and Agriculture Organization (FAO), 1951, 1958, 1963, 1968. *Production Yearbook* (Rome: FAO).

Food and Agriculture Organization (FAO), 1951, 1958, 1969. *Yearbook on Food and Agricultural Statistics,* 1950, 1957, 1968 (Rome: FAO).

Goldberg, E. D., 1970. Atmospheric transport, background paper prepared for SCEP (unpublished).

Gress, F. Reproductive success of the brown pelicans on Anacapa Island in 1970, *Transactions of the San Diego Natural History Society,* forthcoming.

Grolleau, G., and Giban, J., 1966. Toxicity of certain seed dressings to game birds and theoretical risks of poisoning, *Journal of Applied Ecology, 3.*

Hampson, G. R., and Sanders, H. L., 1969. Local oil spill, *Oceanus, 15.*

Hasler, A. D., 1969. Cultural eutrophication is reversible, *Bioscience, 19.*

Heath, R. G., Spann, J. W., and Kreitzen, J. F., 1969. Marked DDE impairment of mallard reproduction in controlled studies, *Nature, 224.*

Herman, S. G., Kirven, M. N., and Risebrough, R. W., 1970. Pollutants and raptor populations in California (unpublished).

Hickey, J. J., and Anderson, D. W., 1968. Chlorinated hydrocarbons and eggshell changes in raptorial and fish eating birds, *Science, 162.*

Hoffman and Henderson, 1967. The fight against insects, *Yearbook of Agriculture* (Washington, D.C.: U.S. Department of Agriculture).

Holme, R. A., 1969. Effects of *Torrey Canyon* pollution on marine life, *Oil on the Sea,* edited by D. P. Hoult (New York: Plenum Press).

Holmes, R. W., 1969. The Santa Barbara oil spill, *Oil on the Sea,* edited by D. P. Hoult (New York: Plenum Press).

Irukayama, K., 1967. The pollution of Minimata Bay and Minimata disease, *Advances in Water Pollution Research, 3.*

Jefferies, D. J., 1969. Induction of apparent hyperthyroidism in birds fed DDT, *Nature, 222.*

Jefferies, D. J., and French, M. C., 1969. Avian thyroid: effect of p,p'-DDT on size and activity, *Science, 166.*

Jefferies, D. J., and Prestt, I., 1966. Post-mortems of peregrines and lannens with particular reference to organochlorine residues, *British Birds, 59.*

Jehl, J. Shell thinning in eggs of the brown pelicans of western Baja California, *Transactions of the San Diego Natural History Society,* forthcoming.

Ketchum, B. H., 1969. Eutrophication of estuaries, *Eutrophication: Causes, Consequences, Correctives*, edited by G. Rohlich (Washington, D.C.: National Academy of Sciences).

Lofroth, G., and Duffy, M. E., 1969. Birds give warning, *Environment, 11*.

McHugh, J. L., 1967. Estuarine nekton, *Estuaries*, edited by G. H. Lauff (Washington, D.C.: American Association for the Advancement of Science).

Manigold, D. B., and Schulze, J. A., 1969. Pesticides in selected western streams —a progress report, *Pesticides Monitoring Journal, 3*.

Matsumura, F., and Patil, K. C., 1969. Adenosine Triphosphatase sensitive to DDT in synapses of rat brains, *Science, 166*.

Menzel, D. W., Anderson, J., and Randtke, A., 1970a. Marine phytoplankton vary in their response to chlorinated hydrocarbons, *Science, 167*.

Menzel, D. W., Anderson, J., and Randtke, A., 1970b. The susceptibility of two species of zoo plankton to DDT (unpublished).

Morris, R. F., and Miller, C. A., 1954. The development of a life table for the spruce budworm, *Canadian Journal of Zoology, 32*.

Muto, S., and Suzuki, T., 1967. Analytical results of residual mercury in the Japanese Storks, *ciconia ciconia boyciana swinhoe*, which died at Okama and Toyooka regions, *Japanese Journal of Applied Entomology and Zoology*.

Nace, R. L., 1967. Water resources: a global problem with local roots, *Environmental Science and Technology, 1*.

Nash, R. G., and Beal, M. L., 1970. Chlorinated hydrocarbon insecticides: root uptake versus vapor contamination of soybean foliage, *Science, 168*.

National Air Pollution Control Administration (NAPCA), 1970. *Nationwide Inventory of Air Pollutant Emissions* (Raleigh, North Carolina: NAPCA).

Nimmo, D. R., Wilson, A. J., Jr., and Blackman, R. R., 1970. Localization of DDT in the body organs of pink and white shrimp, forthcoming.

Olson, J. S., 1970. Carbon cycle and temperate woodlands, *Analysis of Temperate Forest Ecosystems*, edited by D. E. Reichle (New York: Springer-Verlag).

Peakall, D. B., 1970. p,p'-DDT: effect on calcium metabolism and concentration of estrodol in the blood, *Science, 168*.

Peterle, T. J., 1969. DDT in Antarctic snow, *Nature, 224*.

Pitelka, F. A., Tomick, P. Q., and Treichel, G. W., 1955. Ecological relations of jaegers and owls as lemming predators near Barrow, Alaska, *Ecological Monograph, 25*.

President's Science Advisory Committee (PSAC), 1967. *The World Food Prob-Health, Education, and Welfare, U.S. Government Printing Office.)

Report of the Secretary's Commission on Pesticides and Their Relationship to Environmental Health, 1969. (Washington, D.C.: U.S. Department of Health, Education, and Welfare, U.S. Government Printing Office).

Ricker, W. E., 1969. Food from the sea, *Resources and Man* (San Francisco: W. H. Freeman and Co.).

Risebrough, R. W., and Anderson, D. W. Pollutants and shell thinning in the brown pelicans, *Transactions of the San Diego Natural History Society*, forthcoming.

Risebrough, R. W., and Coulter, M. C. Chlorinated hydrocarbons and shell thinning in the ashy petrel, *Oceanochoma Lamochroa* (unpublished).

Risebrough, R. W., Davis, J. D., and Anderson, D. W. Effects of various chlorinated hydrocarbons on birds, forthcoming.

Risebrough, R. W., Huggett, R. J., Griffin, J. J., and Goldberg, E. D., 1968. Pesticides: transatlantic movement in the northeast trades, *Science, 159.*

Risebrough, R. W., Menzel, P. B., Martin, D. J., and Olcott, H. S. DDT Residues in pacific marine fish, *Pesticides Monitoring Journal,* forthcoming.

Risebrough, R. W., Sibley, F. C., and Kirven, M. N. Reproductive success of the brown pelicans on Anacapa Island in 1969, *Transactions of the San Diego Natural History Society,* forthcoming.

Rohlich, G., ed., 1969. Eutrophication: Causes, Consequences, Correctives (Washington, D.C.: National Academy of Sciences).

Ryther, J., 1969. Photosynthesis and fish production in the sea, *Science, 166.*

Schreiber, R. W. Pollutants and shell-thinning of eggs of brown pelicans in Florida, *Transactions of the San Diego Natural History Society,* forthcoming.

Schwartz, H. L., Kosyreff, W., Surks, M., and Oppenheimer, J. H., 1969. Increased deiodination of L-thyroxine and L-trilodothyronine by liver microsomes from rats treated with phenobarbital, *Nature, 221.*

Seba, D., and Cochrane, E., 1969. Surface slicks as a concentrator of pesticides, *Pesticide Monitoring Journal.*

Skou, J. C., 1965. Enzymatic basis for active transport of Na^+ and K^+ across cell membrane, *Physiological Reviews, 45.*

Stockinger, H. E., 1963. Mercury Hg^{237}, *Industrial Hygiene and Toxicology,* Vol. II, edited by F. A. Patty (New York: Interscience).

Sverdrup, H. V., Johnson, M. W., and Fleming, R. H., 1942. *The Oceans* (New York: Prentice-Hall).

Tarrant, K. B., and Tatton, J., 1968. Organo-pesticides in rainwater in the British Isles, *Nature, 219.*

United Nations. *Statistical Yearbook,* 1967 (New York: Statistical Office of the United Nations), 1968.

Verduin, J., 1966. Eutrophication and agriculture (paper presented at American Association for the Advancement of Science Symposium, Washington, D.C.).

Vollenweider, 1969. Eutrophication (Panis: OECD Mimeograph Report).

Weibel, S. R., 1969. Urban drainage as a factor in eutrophication, *Eutrophication: Causes, Consequences, Correctives,* edited by G. Rohlich (Washington, D.C.: National Academy of Sciences).

Whittaker, R. H., and Likens, G. E. *1961 Woodland Forest Working Group of International Biological Program* (unpublished).

Wolman, A. A., and Wilson, A. J., Jr., 1970. Occurrence of pesticides in whales, *Pesticide Monitoring Journal, 4.*

Wurster, C. F., 1968. DDT reduces photosynthesis by marine phytoplankton, *Science, 159.*

Yates, M. L., Holswade, W., and Higer, A. L., 1970. Pesticide residues in hydrobiological environments, *Water, Air and Waste Chemistry Section of the American Chemical Society Abstracts.*

3.
Work Group on Monitoring

Chairman
G. D. Robinson
CENTER FOR THE ENVIRONMENT
AND MAN, INC.

A. P. Altshuller
NATIONAL AIR POLLUTION CON-
TROL ADMINISTRATION
Richard D. Cadle
NATIONAL CENTER FOR ATMOS-
PHERIC RESEARCH
Robert Citron
THE SMITHSONIAN INSTITUTION
Seymour Edelberg
MASSACHUSETTS INSTITUTE OF
TECHNOLOGY
Gifford Ewing
WOODS HOLE OCEANOGRAPHIC IN-
STITUTION
Dale W. Jenkins
THE SMITHSONIAN INSTITUTION
Jules Lehmann
NATIONAL AERONAUTICS AND
SPACE ADMINISTRATION
Henry Reichle
NATIONAL AERONAUTICS AND
SPACE ADMINISTRATION
Morris Tepper
NATIONAL AERONAUTICS AND
SPACE ADMINISTRATION

Rapporteur
Jonathan Marks
HARVARD LAW SCHOOL

3.1

The Concept of Monitoring

When the miner's canary died, it was time to get out of the mine. The canary "monitored" the mine air and gave an indication of potential disaster due to odorless, invisible methane. The immediate action necessary was clear; long-term solutions could be considered later.

But when we are concerned with a global environmental problem, this type of monitoring is insufficient. Because we cannot escape from the earth, we must have more than a sentinel to sound an alarm if a critical threshold is passed; we must know what it is that kills our "canary," where it comes from, and how to turn it off at the source.

Accordingly, we think "monitoring" is best conceived of as systematic observations of parameters related to a specific problem, designed to provide information on the characteristics of the problem and their changes with time. The parameters and problems with which we have been concerned are those of the global environment. And though any monitoring program will provide information useful to dealing with local and regional problems, our concern has been with identifying existing and potential monitoring systems capable of securing the information necessary to deal with the critical global problems identified by the Study of Critical Environmental Problems (SCEP).

For every one of the global problems that have been identified, we find we have insufficient knowledge of either the workings or the present state of the environmental system (see reports of Work Groups 1 and 2). This hinders us as we attempt to design monitoring that will not only warn us of change but also provide information upon which we can base rational and efficient remedial action. In most instances we can suggest a likely analogue of the canary, but we do not know what action would be best once our bird shows up sick. Further, we are persuaded by our colleagues that global systems both physical and biological are so complex that the ultimate consequences of any disturbance cannot at present be predicted with confidence.

For these reasons our report is concerned not only with monitoring in its sense of providing warning of critical changes but also with measurements of the present state of the system (the "base line") and with measurements in support of research into the workings of the system. We mention the need for this research where it is apparent to us; we have not attempted to provide a complete assessment of research needs. In general, however, we have agreed that research is most needed in providing a closer specification of the present state of the planet and in developing a more complete understanding of the mechanisms of interaction between atmosphere, ocean, and ecosystem.

3.2
Monitoring Techniques and Systems

Before we turn to an analysis of the monitoring aspects of the critical global environmental problems identified by SCEP, we feel it necessary to look more generally at the current state of monitoring techniques and systems, both to provide a framework within which the later recommendations can be seen and to point out areas on which emphasis should be placed as the effort to obtain information for environmental management continues.

3.2.1
Economic and Statistical Monitoring

If we are concerned with predicting the accumulation of a pollutant in the environment, its rate of input must be known. For example, to evaluate the global CO_2 problem, we require a long-term prediction of the total atmospheric content, which, in turn, depends in part on statistical projections of fuel production and consumption. To evaluate the contribution of SO_2 to the global particulate problem, we require projections of natural, industrial, and energy-conversion emissions of SO_2 and SO_2-precursors which take into account possible control and abatement measures.

If we are concerned with evaluating the effects of alternative control technologies on pollutant levels, we need quantitative information about the flow of materials which will be altered by control technology to include inputs, wastes, and end products at each stage of the process (Ayres and Kneese, 1968).

Both these kinds of activities seem to us to be essential forms of monitoring. To some degree they are already being carried out, for example, in industry as a part of the management process, and in government as a part of economic policy making and of already existing regulatory activities.

Yet, the focus of the gathering and synthesis of data concerning industrial, agricultural, domestic, and energy-producing activities has not commonly included an "environmental effects" component. Moreover, research into the natural pathways and degradation of pollutants once they are deposited in the environment has as yet yielded little quantitative data. Nor is our knowledge of the functioning of particular ecosystems sufficient to allow us to quantify effects of a particular pollutant when it, for example, eradicates one or a group of species within the ecosystem. This has meant that we lack information on which to base projections and models for decision making. It has also meant that the organization of data on which to base an analysis of the global environmental problems considered by this conference has been a tedious, often approximate, and sometimes impossible task.

New methods for gathering and organizing economic and statistical data must be developed if we are to have the "handle" we need to deal with environmental problems. New centralized collection and collation points are needed. Federal regulatory bodies in the United States, such as the National Air Pollution Control Administration (NAPCA), have begun this task for their own areas of responsibility. But there exists no effective organization summing data across traditional areas of environmental responsibility, such as air and water pollution. Nor do we have any comparable international organization or, indeed, any effective standards to ensure, for example, that the industrial production data collection going on across the world will be of comparable precision and focus. All these tasks must be accomplished if we are to use effectively our economic and statistical monitoring potential.

Recommendations

1. We recommend the development of new methods for gathering and compiling global economic and statistical information, which organize data across traditional areas of environmental responsibility, such as air and water pollution.

2. We recommend the propagation of uniform data-collection standards to ensure, for example, that industrial production data

collection being carried out across the world will be of comparable precision and focus.

3.2.2
Physical and Chemical Monitoring

Physical and chemical monitoring methods are used to determine the amount of a contaminant in a sample of soil, water, air, or organism. Physical methods are also used to determine a property of an environmental system as a whole, such as the refractive index or the albedo of the atmosphere.

There are numerous examples of this type of monitoring. NAPCA's network of sampling stations monitors the quality of urban air. Weather satellites monitor the formation of hurricanes. The essence of good monitoring of this type is to measure what is needed, and no more, with the precision that is needed, and no more, and to maintain standards indefinitely.

Traditionally, monitoring of this type is carried out in networks of fixed stations. The entire operation may be completed at these stations or a sample may be taken to a central laboratory for examination or analysis. In either case, central coordination of methods and central standardization is necessary. Monitoring is now extended to measurements on ships, aircraft, and satellites. Moreover, it has become an international activity, and international coordination of standards is necessary.

Measurements of solar radiation at the ground form a good if little-known example. They are made in numerous countries by national meteorological services, universities, agricultural research stations, and so forth. They are also collected, edited, and published at a Soviet observatory under arrangements guided by the World Meteorological Organization. These measurements are not of uniform quality, but the best which can be identified have been standardized by instruments compared internationally on an ad hoc basis. The last major comparison was arranged bilaterally between an American manufacturer and a Soviet university. The necessary nominal institutional anchor for these comparisons is the Radiation Commission of the International Association of Meteorology and Atmospheric Physics within the framework of the International Council of Scientific Unions. To give another example, our knowledge of the CO_2 content of the atmosphere is due to the interest, skill, and cooperation of small groups of scientists in the United States and Sweden (Pales and Keeling, 1965; Keeling, 1970).

Recommendation

We recommend the development and expansion of current physical and chemical monitoring systems and techniques. Specific recommendations are included in section 3.4 on "Monitoring Critical Global Problems."

3.2.3

Biological Monitoring

Even though our interest in environmental pollution stems from our concern about its effects on living organisms, the concept of using such organisms, either individually or as a population or species, as tools to monitor the state of the environment is still a relatively untested one. Moreover, although the study of natural ecosystems has long been an important scientific activity, the observation and evaluation of changes in these finely tuned systems have not yet been systematized to yield warnings about harmful contaminants.

Yet living organisms can serve as excellent quantitive as well as qualitative indices of the pollution of the environment. Plants and animals are continually exposed and can act as long-term monitors that integrate all environmental effects to reflect the total state of their environmental milieu. They can show the pathways and points of accumulation of pollutants and toxicants in ecological systems. Their use can remove the extremely difficult task of relating physical and chemical measurements to biological effects.

There are numerous examples of the use of biological organisms as monitors. The miner's canary has already been mentioned. Rats are being used today as air pollution monitors, for CO. There is currently a widespread and effective pesticide monitoring network in the United States based on the analysis of biological material (Murray et al., 1970).

Yet further planning and coordination is necessary if biological monitoring is to play its optimum role in environmental information gathering. The pesticide monitoring network, already noted, could have by now made a greater contribution to our understanding of this critical problem had it been initially combined with air and soil monitoring into a comprehensive program for dealing with the problem of pesticide pollution instead of having been implemented as a valuable but inadequate response to the discovery of *symptoms* of pesticide poisoning in animals.

Planning for biological monitoring of global environmental problems should consider three basic methods: (1) international networks of biological surveys; (2) international networks of terrestrial and aquatic ecological base-line monitoring stations; and (3) the use of biological organisms as sentinels, detectors, and indicators and for biological assay of man-produced contaminants and changes.

1. Biological surveys are broad geographical evaluations of the population size, reproductive success, and health of species of living organisms. Such surveys can be accomplished by the census of organisms during such critical periods in the life cycle as breeding, nesting, overwintering, migration, and flowering. Commercial and other harvest data on organisms such as fish and game birds are already available examples of information that could make up part of such a survey. Production figures from plant-food crops and lumber can also be used, provided information about the growing cycle and treatment with fertilizers and pesticides is known so that the data can be analyzed.

2. A global network of ecological base-line monitoring stations would show general changes of major significance to the flora and fauna of the world. A complete system would require establishing stations in each of the following biomes: deciduous forests, conifer forests, tropical forests, savannas, grasslands, deserts, tundras, estuaries, and various ocean areas. The terrestrial base-line stations would enclose permanent, protected natural areas with a long record of ecological and meteorological study. The site would best be at least 4 kilometers square with a large surrounding area relatively free from human intervention and direct contamination. A meteorological station and a contaminant sampling station would be located near the base-line station to permit correlation of biological results with physical and

chemical measurements. Comparable samplings of temperature, rainfall, solar radiation, and the composition of the atmosphere would be made near aquatic stations.

Within the area of the base-line station detailed ecological studies would be combined with biological surveys of the kind just described. The results obtained in these stations would be compared with those obtained at impact stations located to study the direct biological effects of man, especially those caused by modern agricultural practices, cities, and industry.

3. Many species of biological organisms have great sensitivity to pollutants. Many animals and plants can be used as "early warning" sentinels for particular pollutants. Others can be used to provide graphic records of pollutants; by looking at certain plants, for example, it is possible not only to identify the presence of certain pollutants in a given area but also to gain information on this concentration and its variation in time. Other organisms, simply by their presence in a particular place, signal that a particular pollutant is also present; the proliferation of algae in lakes, for example, indicates an excess of nutrients like phosphorus.

Recommendations

1. We recommend early implementation of a set of ecological base-line stations in remote areas that would provide both specific monitoring of the effects of known problems and warnings of unsuspected effects.

2. We recommend central coordination and, where necessary, modification of national and regional surveys of critical populations of fish, birds, and mammals from commercial catches, harvests, and surveys. This would provide an early warning system by monitoring highly sensitive and vulnerable species.

3.2.4
Modern Technology and Monitoring

A major problem in those monitoring systems that require precise measurement and rigorous attention to detail is to combine the continuity and security of institutional control with the continued devotion of the scientists and skilled technicians who are often required to work in isolated regions. Further, the extreme care that is necessary in sample handling makes the standardization and interpretation problem extremely difficult when measurements must be made independently at each one of many stations within a global network (*in situ* monitoring) (Pales and Keeling, 1965).

For these and other reasons we must continue to investigate the possibility of monitoring remotely, by automatically reporting or man-operated instruments carried in earth satellites or airplanes.

The unmanned earth satellite, especially, has many virtues as a vehicle for monitoring equipment. It can provide global coverage in very short periods of time. It can carry equipment without subjecting it to environmental stress. If large amounts of information from widely distributed points are required or if a particular monitoring system can be "piggybacked" on a satellite with other functions, there may be cost advantages. *In situ* techniques in general yield data whose precision has not yet been equaled by remote techniques for measuring the same parameter. The techniques now available for use in satellites must be improved if they are to provide the high-precision data that are required.

It must be pointed out that interesting and useful data have already been obtained from satellites for use in the atmospheric and oceanographic sciences. This, combined with preliminary information about satellite monitoring experiments planned in the next few years, indicates that satellite solutions to many environmental data-gathering problems could be available in the not-too-distant future.

What is clearly needed is a series of evaluations of appropriate satellite techniques, including both scientific feasibility studies and cost-benefit analyses. Once scientific feasibility is established, the aim of such an evaluation would be to determine the optimum system for providing the required information, be it a satellite, ground, or mixed system.

In addition to this comparative evaluation of remote and *in situ* monitoring, there is need for continuing research into monitoring per se. In recent years new and powerful techniques have been developed in radiometry, radar, and spectroscopy which are operated in the microwave, infrared, and visible regions of the electromagnetic spectrum. Furthermore, digital data-processing systems have been developed which permit rapid handling and analysis of large data flows. There is a need for continuing research into these systems, as well as into the integration of diverse requirements to ensure optimum use of resources, and into scientific and administrative coordination.

Recommendations

1.   We recommend generally a series of evaluations of appropriate satellite measurement and monitoring techniques, including both scientific feasibility studies and cost-benefit analyses, aimed at determining the role of satellites in an optimum monitoring system for the problems dealt with by SCEP.

2.   We recommend continuing research into the application to environmental monitoring of radiometry, radar, spectroscopy, digital data processing and other newly available techniques.

3.3

The Precursor of Monitoring: An Ocean Base-Line Sampling Program

Throughout this study we have often failed to discover enough existing data either to allow us to say, "This is definitely a problem," or to provide a starting point for the solution of an identified problem.

We do not know, for example, the concentration gradients of CO_2 in its ocean reservoir. We do not know the natural production of SO_2. We need more information about the concentration of DDT in its atmosphere, soil, and river routes to the sea.

To obtain these data, both one-time and continuing surveys are needed; these surveys will help us establish a base line for analysis. We have already briefly discussed a system of possible ecological base-line stations. We shall discuss now the kinds of base-line atmospheric data that are already available or that can be obtained as a part of what we see as clearly necessary atmospheric monitoring programs. In this section we present a comprehensive base-line survey program for the oceans, developed by members of a SCEP group that considered coastal and oceanic problems.* It is meant to serve both as an example of the kind of one-time survey this Work Group thinks is necessary and as a practical proposal for immediate implementation.

A one-year program is proposed to collect approximately one thousand samples from the following components of the environment: oceans, organisms, rivers, glacers, rain, and sediments.

The program has two basic objectives: (1) to provide a comprehensive-historical record of the current state of the ocean system, primarily focused on the concentrations of pollutants in their routes to the ocean and reservoirs in the ocean; (2) to identify substances among those which were sampled and analyzed whose concentrations are such that they should be examined further. The first objective will provide a point of comparison for other general surveys of the ocean system, which should be undertaken periodically in order to check for major changes in the system. The second objective, if further toxicity studies support a conclusion that a particular concentration poses a threat to man, will lead to a continuing monitoring program.

The choice of a thousand samples, each to be analyzed for a broad spectrum of contaminants, seems the minimum necessary to provide a base broad enough to allow analytical conclusions that are statistically significant for the ocean system as a whole.

3.3.1

The Oceans

Since the pollutants of potential importance in the ocean may take a variety of forms (gases, solutes, particulates, oils) and may enter through various paths (atmosphere, rivers, ships), it is suggested that a base-line set of water samples must be spaced both horizontally and vertically in such a way as to detect all of these forms. Both the physical circulation of the ocean and the distributions effected by biological transport must be taken into account so that a proper framework for assessment of the ultimate distributions in the oceans, by whatever paths and transformations, can be established.

A tentative framework for base-line sampling of ocean water includes the following:

---

* The members of this group were Geirmunder Arnason, Edward D. Goldberg, M. Grant Gross, Frank G. Lowman, and Joseph Reid.

Vertical: Three kinds of vertical spacing are suggested: (1) a sample of surface film by itself; (2) a shallow array including a sample of surface film, mixed layer, and pycnocline;* and (3) a deep array combining (2) with samples from depths of about 1,000 meters, 2,000 meters, 3,000 meters, and near the bottom.

Deep array: Samples should be taken in the centers of the major gyres† and in the major mediterranean seas (Arctic, Gulf of Mexico, Mediterranean, Okhotsk, Bering, and Sea of Japan). With a central equatorial array for each ocean the total is 20 deep arrays of 7 samples each, or 140 samples.

Shallow arrays: Samples should be taken in the major eastern and western boundary currents (about 16 positions) and at the eastern and western edges of the three oceans at the equator. This totals 22 arrays, with 3 samples each, or 66 samples.

Surface film: In addition to the 42 surface film samples taken as part of shallow array sampling, an additional set of perhaps 50 should be collected, spaced generally over the ocean but with emphasis on the major zones of precipitation and evaporation, and on the major shipping routes (especially oil routes).

3.3.2

Organisms

A base-line measurement of the amounts of man-made pollutants in the marine biosphere must consider the following in establishing a sampling program: (1) geographical distribution of primary biological productivity; (2) geographical distribution of world fisheries; (3) sites of major river outflows; (4) general air circulation patterns; (5) sites of major down- and upwellings of water; (6) sites of centripetal centers of major water gyres; (7) desired coverage of marine populations of organisms; (8) ease of sampling; (9) cost.

The collection of adequate samples of marine organisms is especially important in areas of high primary productivity and major fisheries since several of these areas are (1) near sites of dense population of humans and the sites of pollution production; (2) support relatively large marine biomasses which may serve as reservoirs with slow turnover rates for pollutants; and

---

* Pycnocline: A region of very rapid increase with depth of water density; an effective barrier to vertical mixing.

† Major gyre: A system of near-surface currents, with the appearance of a closed circulation, on the scale of a major ocean basin.

(3) provide food pathways leading to man through man's utilization of marine fishery products.

Sampling stations shown in Figure 3.1 are located on continental shelves at sites of high productivity, major fisheries, the outflows of major rivers, and at the sites of down- and upwellings of water. In addition, sampling stations are placed in the Mediterranean Sea and off the southern shores of Africa, Australia, and South America. The present plan calls for the collection of 42 composite samples each of demersal fish, mollusks, and pelagic fish and 12 composite samples of benthic crustacea—a total of 138 samples.

Except for two sampling stations in the Central Pacific, downwind from the upwelling area off the Pacific Coast of South America, the central oceanic areas are not samples. The great average depth of these sites precludes sampling of demersal organisms; the areas contain greatly reduced and sporadically occurring populations of pelagic fishes. Two types of samples should be collected from the central oceanic regions, to include net-plankton and flying fish. Plankton may be taken on a predetermined collection grid, as shown in Figure 3.2. Sufficient flying fish normally come aboard research vessels operating in these regions at night to supply adequate samples for the central ocean areas. A total of 142 plankton samples and 40 flying fish (or more, if available) should be collected from the Atlantic, Pacific, and Indian Oceans and the Mediterranean Sea.

In summary, the program suggested would consist of the following samples:

Fish	124
Mollusks	42
Crustacea	12
Plankton	142
Total	320

3.3.3

Rivers

Rivers are important routes by which numerous wastes reach coastal oceans. Identification of major routes and reservoirs requires that river water, riverborne sediment, and related sediment deposits be samples during base-line studies. For the initial study, a limited number of areas are suggested.

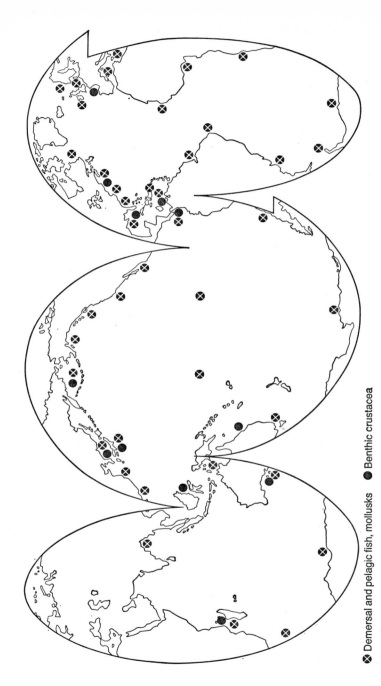

⊗ Demersal and pelagic fish, mollusks   ● Benthic crustacea

Figure 3.1   Fish, Mollusk, and Crustacea Sampling Stations

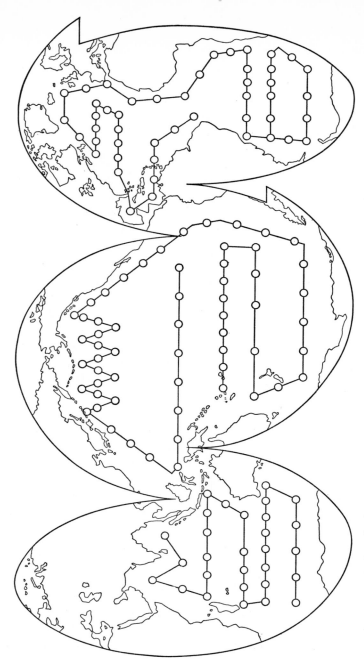

Figure 3.2 Net Plankton Stations

River water and sediment samples should be collected in volumes proportional to river flow and composited over a 6-month period. One sampler near the river surface (perhaps on a float) would collect river water. Another near the bottom would collect water and sediment.

Four samples a year from 19 river systems (Table 3.1) total 76 samples per year. The number of samples taken in rivers could profitably be doubled to provide more information about time variability in such systems.

Samples should also be collected from recently deposited sediments at the head of the estuary of each of the rivers listed in Table 3.1. The sample should be typical of riverborne sediment and wastes accumulating in the estuary (the materials most likely to be removed by dredging or washed out by floods).

To understand movement of particulate matter out of a river, it is necessary to study deposits in the adjacent continental shelf or associated delta. Eleven areas (Table 3.2), including four delta systems, are suggested for study. Several are associated with major rivers to be studied. Others (such as the Japan Sea and the two Japanese bays) are selected because of the importance of the regions as developing industrial areas. The limited number of samples (approximately 50) should be composited to permit estimates of waste concentrations in recently deposited sediment.
3.3.4
Glacial, Rain, and Sediment Samples
A historical record, which can be interpreted to determine the rates of atmospheric entry of pollutants to the environment over the past several hundred years, is contained in the permanent snowfields (glaciers), which are so located as to allow an examination of fallout from all of the major wind systems. Time marks can be retrieved from the glacial record by lead-210 geochronologies, firn stratigraphy, or oxygen isotopic stratigraphy. Variations in oxygen isotopic compositions of glacial ice allow the yearly sequences to be subdivided into summer and winter periods.

Table 3.3 lists examples of glaciers that might be assayed. These glaciers and the winds they monitor have in most cases already been studied experimentally as recorders of atmospheric transport.

A sampling program of these six zones is proposed with

Table 3.1
Rivers Included in Base-line Study

North America	Europe	Asia
Hudson	Rhine	Ganges-Brahmaputra
Mississippi	Danube	Yellow (Hwang Ho)
St. Lawrence	Po	Amur
Columbia	Thames	Ob
		Indus
South America	**Africa**	
Orinoco	Nile	
Amazon	Congo	
Plata	Niger	

glacial samples representing 1970 summer and winter, 1969 summer and winter, 1968 summer and winter, 1965, 1960, 1953, 1943, 1930, 1900, and 1800.

For the trades and westerlies in the Northern and Southern Hemispheres, composite rain samples (perhaps a year's collection of rainwater) from single locations in both the Pacific and Atlantic Oceans would be most useful to compare with the glacial results as well as to give a measure of precipitation washout.

The total number of samples proposed:

78 permanent snowfields

48 rain samples

126

Further historical information is contained in the deep sea sedimentary column that accumulates at extremely slow rates, fractions of a millimeter to centimeters per thousand years. These deposits are composed of rock debris from the continents, volcanic materials and their degradation products, oceanic precipitates such as the ferromanganese minerals and barite, and animal and plant frustules.

Various disturbances reduce the utility of an extensive sampling program, but the biological remains, calcareous and siliceous tests, do provide a measure of the removal of materials from surface waters to the sediments by biological agencies. A suite of 4 samples each, of siliceous and calcareous ooze, from the Pa-

Table 3.2
Continental Shelf Areas, Deltas, and Harbors

Osaka Bay	Baltic Sea
Tokyo Bay	Black Sea
Ganges-Brahmaputra Delta	Japan Sea
Mississippi Delta	North Sea
Nile Delta	Yellow Sea
Rhone Delta	

cific, Indian, and Atlantic Oceans can provide a most reasonable set of base-line materials.

Recommendations

1.   We recommend the implementation of this 1,000-sample survey of the oceans to provide general knowledge of the distribution of man-made influences in the oceans.

Without information supplied by such a survey we shall have great difficulty in specifying the volume and distribution of observations necessary to monitor for critical environmental problems. We note that information of the same kind must also be obtained from living matter (see Recommendations in Section 3.2.3) and from the atmosphere (see introduction to section 3.4.2).

2.   We recommend, either as a part of the foregoing survey or independently, examination of glaciers and sediments to help remedy the current lack of an adequate historical record of the oceans and of world climate and, especially, to clarify at least the recent variations of atmospheric and oceanic particulate content.

3.4

Monitoring Critical Global Problems

Our conception of monitoring associates the activity with identified problems. The deliberations of SCEP appear to us to have isolated the following as currently critical global problems:

1.   Ecological effects of DDT and other toxic persistent chlorinated hydrocarbons.

Table 3.3
Glaciers That Might Be Assayed

Glacier	Wind System
Greenland	Northern Hemisphere polar easterlies
Yukon Territory or Mount Olympus, Washington	Northern Hemisphere westerlies
Mt. Orizaba or Mt. Popocatepetl, Mexico	Northern Hemisphere trades
Andean Glaciers	Southern Hemisphere trades
Tasman Glaciers, New Zealand	Southern Hemisphere westerlies
Antarctica	Southern Hemisphere polar easterlies

2.   Ecological effects of mercury and other toxic heavy metals.

3.   Climatic effects of increasing carbon dioxide content of the atmosphere.

4.   Climatic effects of the particle load of the atmosphere.

5.   Climatic effects of contamination of the troposphere and stratosphere by subsonic and supersonic transport aircraft.

6.   Ecological effects of petroleum oil in the oceans.

7.   Ecological and climatic effects of physical and thermal changes at the earth's surface, including changes in land use.

8.   Ecological effects of nutrients in estuaries, lakes, and rivers.

In this section we consider in turn the nature of the monitoring effort associated with each of these problems.

Where we feel competent, we have added estimates of the cost of some parts of the effort. We add the caveat that these estimations are based on our visualization of how the measurements might be made. Furthermore, time did not permit the rationalization of estimates made by different members of the group. If monitoring schemes are activated, others will doubtless activate them, and they will differ in some degree, perhaps considerably, from our visualization. The basis of costing will therefore also differ.

We add a second caveat that is even more weighty. It would be far better to monitor global problems by using an integrated system than to tackle them piecemeal. To some degree we have assumed integrated systems in making our cost estimates and in estimating the number of sampling stations and the kind of data processing required. Yet because of the lack of base-line data and the nature of the SCEP exercise, it has been impossible for us to do much more than state the need for a study of the components of an optimum integrated monitoring system for global problems.

3.4.1
Monitoring Chlorinated Hydrocarbons and
Toxic Heavy Metals

We here group together two very different classes of materials used by man in very different activities. But the substances are closely allied from the aspect of monitoring. Their threat to the environment—the poisoning of living things other than those which man, mistakenly or not, wants to poison—is the same. Their routes and reservoirs in the environment are the same. It happens also that the required monitoring technique, quantitative chemical analysis of trace substances, is similar in the two cases. Hence, at least in general, we can and should monitor the substances in the same places and with the same degree of detail.

We are dealing here with persistent poisons, which man has found cause to use either because they kill pests or are important in industrial processes. The chlorinated hydrocarbons not only kill target insect pests but also many beneficial biological control organisms. They are accumulated in harmful amounts in the higher levels of various food chains.

Lead, mercury, and certain other heavy metals can be harmful to men and animals if inhaled. They may be harmful to plants and animals in streams, lakes, and oceans. When they reach the earth's surface, by the scrubbing action of rain, for example, they may have harmful effects on plants. The heavy metals are introduced into the air as particles by automobiles, many industries, and by power plants. They are a part of the waste effluents dumped into rivers by many industrial processes.

The chlorinated hydrocarbons, since they include such a wide variety of products, are put into the environment by several different routes. In the application of pesticides to agricultural lands, a significant proportion is lost directly to the air, from which it later falls in rain; another portion is washed out of the soil into watercourses; another portion remains in the soil and is degraded there. Some chlorinated hydrocarbons ultimately accumulate in the oceans.

The knowledge from which the above descriptions of routes and reservoirs is taken is primarily qualitative; what quantitative data we have is generally rough, with the exception of some world production and use figures for the chlorinated hydrocarbons, United States emission figures for the heavy metals, and isolated data on the accumulation of such substances in organisms. There is a need for base-line surveys and monitoring in all these areas.

We must discover what has happened to the quantities of chlorinated hydrocarbons and heavy metals we have produced over the past decades. We must map the current flows of such materials in the environment, identifying, for example, how much persistent insecticide is transported in the atmosphere, how much of this is deposited as rain, and where it is deposited.

Our monitoring effort, then, must include the atmosphere, soil, oceans, rivers, and biological material. (For amplification of the foregoing discussion, see reports of Work Groups 2, 5, and 6.)

Compilation of Source, Use, and Dispersal Data

In order to evaluate the importance and size of the problems involved, it is necessary to have the world production and use data, including especially data on dispersal, emission, and direct pollution into the air, water, and soil. We should know the geographical area of dissemination into the environment as well as the volume and dates of dispersal.

At present, statistics on production and use are available for most of the world for both the chlorinated hydrocarbons and heavy metals (the Soviet Bloc data are lacking). Few accurate U.S. and world data are available on the escape of heavy metals to the environment.

Monitoring for Chlorinated Hydrocarbons and
Toxic Heavy Metals

SAMPLING AND ANALYSIS. The techniques for collecting air, soil, and water samples for chlorinated hydrocarbon and heavy metal analysis are well developed and are routinely carried on in a number of laboratories. (Feltz, 1969, National Air Pollution Control Administration [NAPCA], 1969; Cadle, 1970; Carver, 1970; Murray, 1970; Sand, 1970.)

Air. Most of the organic chlorides of concern have low but significant volatility and tend to be associated with airborne particles at the relatively high concentrations at which measurements are generally made. Since the particles have a short residence time in the air (probably a few days), the organic chlorides probably have a short residence time as well. Accordingly, a fairly large number of collection points seem necessary to assure adequate sampling.

The analytical methods employed assume for the most part that the chlorinated hydrocarbons are associated with airborne particles and start by the collection of such particles on a filter. However, when the chlorinated hydrocarbon concentrations are very low, as in the ambient atmosphere, this assumption may lead to gross errors since the chlorinated hydrocarbons may be largely in the gas phase. Thus, it is recommended that at least at first the atmosphere should be analyzed for both the gas and particulate phases of these substances. The usual analytical technique involves the collection of particles on a filter, extraction of the filter, and a chromatographic analysis of the extract. Special techniques may have to be developed for gas-phase analysis.

Since the heavy metals are associated with atmospheric particles, their residence time in the atmosphere is relatively short, probably a half-life of a few days. Thus, a large number of sampling points is advisable, of the same order as for the organic hydrocarbons. The U.S. Public Health Service already monitors a large number of metals routinely through its National Air Sampling Network. Many of these 185 urban and 51 nonurban stations could provide the needed samples (NAPCA, 1969).

The particles containing the metals can be collected by filtration and analyzed in a central laboratory. Probably the same collection system as described for sulfates later in this report will suffice. Probably 10 to 15 metals would be monitored, certainly including lead, mercury, copper, zinc, arsenic, cadmium, and vanadium. Determination of the concentrations of this number of metals is relatively easy once samples have been obtained and prepared for analysis.

Water and soil samples. Samples can be collected in prepared containers from lakes and rivers with a depth-integrating sampler. Bottom sediments from rivers can readily be obtained with either a bed material sampler or a piston core sampler containing a Teflon liner.

Collections of samples in estuaries and open ocean locations can be done by use of collectors presently employed in oceanographic activities. Water samples of about 200 liters can be obtained by large volume samplers.

Pollution in sediments accumulated from water runoff from the land usually is deposited on the bottom surface and may be a good index of general pollution levels in the local hydrologic environment. Both the piston-type bed material sampler and commercial core samplers are adequate for collecting sediment samples.

Soil samples should consist of cores 2 inches in diameter and 3 inches deep.

Required sensitivity for detecting trace amounts of chlorinated hydrocarbons in water, suspended particles, bottom sediments, atmospheric dryfall, and biota is 5 ppt. The electron capture gas chromatograph is needed for minimum identification. Microcoulometry, coupled with gas chromatography, may be used for confirmation when concentrations permit. Analysis is performed with extracts injected into two columns having different retention properties. Positive identification can usually be obtained by corroboration of results using at least two types of gas chromatography columns. Additional confirmation when required can be made using specific detectors such as chloride, sulfur, or nitrogen microcoulometry and gas chromatography phosphorus detectors. Analysis by a third column with different retention times can be helpful, and, if infrared spectroscopy is available, special microtechniques applied to separated materials can aid in positive identification.

SAMPLING NETWORK. In order to obtain systematic sampling of all significant air, water, land, and biota systems, a network of some 200 integrated sites strategically located around the world seems minimally necessary (or a larger number if some are specialized, for example, sampling only the atmosphere). To set exact numbers and locations, more information of the kind that would be obtained through the ocean base-line sampling program proposed in section 3.3 is needed.

In general, however, each land-monitoring site and the sites selected for snow and ice fields would be required to sample dryfall, rainfall, ground surface cores, and land plants on a periodic basis, perhaps monthly. Dryfall and rainfall sampling would require multiple samplers to insure that individual specimens are obtained for chlorinated hydrocarbon and heavy metal analysis. The lake, river, estuary, continental shelf, and deep ocean sites would be required to sample water surface film, suspended particulates, the water itself, and marine biota, perhaps monthly. All sites but the deep ocean station would obtain bottom sediment samples somewhat less frequently. Sample frequencies and priorities would obviously change as such a program developed and as base-line information on pollution levels and variability was obtained.

Monitoring of chlorinated hydrocarbons and heavy metals is presently under way in many countries. In trying to set up a program of sampling to provide integrated and systematic information on a global scale, these programs as well as facilities being employed for other monitoring and surveying tasks should be used if at all possible. In fact, a first step toward developing the kind of comprehensive network just described would be the preparation of a survey summarizing and evaluating existing programs and facilities worldwide (for an example of such a survey, see Citron, 1970).

Without this kind of initial survey, it is difficult to estimate costs of a comprehensive monitoring program. In general, capital and operating costs would include the costs of sample collection, sample analysis, operating a central standards laboratory, data processing, and logistical support.

Recommendations

1. We recommend that chlorinated hydrocarbons and toxic heavy metals be monitored with a view to answering two questions:

What has happened to the material that we have used in the environment or allowed to escape into it?

What will happen to material that we use or release, in specified circumstances, in the future?

To provide answers to these questions, we recommend that current monitoring efforts be systematized and extended to provide adequate records of

a. The concentrations of these materials in living things and their effect on life.

b. The production, use, and escape to the environment of these materials.

c. The concentrations of these materials in the soil and air, in precipitation, and in lakes, rivers, and oceans.

2. We recommend a system of about 200 stations that, according to their location, would monitor air content, soil content against depth, water content against depth, dry fallout; washout by rain, bottom sediment content, and content of selected key species.

Our knowledge of existing concentrations and effects is not sufficient to make firm specifications for the system. Hence the number and location of stations should be kept under continuous review.

3.4.2

Atmospheric Composition and Possible Climatic Changes

The possibility has been widely publicized that man, by changing the chemical composition of the atmosphere by burning fossil fuel, might also inadvertently change the physical state and motions of the atmosphere and thus the climate (Shapley, 1953; Callender, 1961; Mitchell, 1969). Elsewhere in this report it is emphasized that the present state of atmospheric science is such that no certain answer can be given to the question, "Can man's activities produce catastrophic changes of climate?"

What is clear to us is that the stake in terms of human welfare is so high that the relatively small investment in monitoring and research should be made. In relation to climate two atmospheric constituents appear to be of importance, carbon dioxide and particles. They present monitoring problems of diverse kinds, but so far as measurements in the atmosphere are concerned they have some things in common. We shall examine these problems after we have looked more generally at atmospheric monitoring.

The physical state of the atmosphere is continuously observed in great detail for general meteorological purposes, so that if we were interested in the distribution of a contaminant and knew where it entered the atmosphere, we could trace its path with useful precision, and estimate the extent to which it spreads by diffusion. If we wished to monitor any pollutant, we could locate sampling stations by reference to our knowledge of atmospheric motions. However, we have now very little knowledge of the distribution within the atmosphere of some pollutants.

Reasoning of this kind leads us to suggest that global monitoring of a trace contaminant or constituent with a long residence time, for example, CO_2, can be accomplished by about 10 stations. Constituents with short residence times with a number of localized sources, for example, some particles, we judge to require about 100 stations (Commission for Air Chemistry and Radioactivity, 1970). Experience for 1 or 2 years with measurements at about 100 stations, including aircraft collection, would allow us to reassess the need, decreasing or increasing the number of stations or relocating some if necessary. In this sense we would acquire an atmospheric "base line." Development of a comprehensive, permanent monitoring system requires these initial measurements.

Prediction of the Carbon Dioxide Content of the Atmosphere

The earth's accessible reservoirs of CO_2 are in the atmosphere, the oceans, the living and decaying biomass. There is continuous exchange between these reservoirs, and man is continuously releasing the CO_2 that is stored in fossil fuels. To allow study of the possibilities of climatic change, we must know what proportion of the released gas remains in the atmosphere, we must be able to extrapolate the trend for many years ahead, and we must be able to update the extrapolation continually. If the existence of a climatic trend were to be firmly established, the obvious corrective actions—curtailment of combustion and initiation of major population movements—would imply change in the whole pattern of human life. We should need a long lead time in order to make the change; monitoring the CO_2 problem therefore calls for the longest possible extrapolation of the average CO_2 concentration in the atmosphere. This estimate must be based on

1. Accurate and reliable data obtained from a monitoring system that senses mean global CO_2 changes, free from perturbations from local sources and sinks.

2. A measurement program that investigates the earth's CO_2 system, including the three interacting components of source, route, and reservoir.

THE PRINCIPAL SOURCES OF CO_2. Possible sources of CO_2 are changes in the biomass (mostly forests), the combustion of fossil fuels by man, and net release or take-up of CO_2 by the oceans. Monitoring, which yields a global map of source position (and intensity at each position), is done by obtaining and analyzing the available data on man's activities such as the burning of fossil fuels, lumbering, or forest clearing for agricultural purposes.

These data need be updated only every 5 years since these sources do not vary significantly in shorter periods of time. Use should be made of the high resolution data available in space photography (Minzner and Oberholtzer, 1970). Use of such data could help determine the characteristics of sources that have not been documented in available records.

There appear to be very few data useful in assessing the effect of man on the carbon cycle in forests. A research program should be instituted to obtain and interpret these data, which would note in a direct manner the net rate of change of carbon that is fixed in the forests.

Data-gathering requirements.

1. Spatial resolution. The production and usage records should be available for each large industrial region and forest area in each country.

2. Temporal resolution. The needed resolution for industrial, domestic, commercial, and transportation combustion is seasonal, yielding annual averages on a global basis. These 12-month averages should be upgraded every 5 years.

3. Accuracy. For monitoring, estimates of source changes should be made with the highest possible accuracy—± 20 percent should certainly be attainable. For research, carbon changes in living trees and woody debris should probably be measured to better than ± 1 ppm.

Data-gathering techniques.

1.   Monitoring. Empirical source models would be developed from available records, augmented by photographs taken in the visible and infrared by satellite systems already in operation and those planned.

2.   Research. The carbon cycle data would be obtained by one or more of the following proved techniques: (a) Harvest techniques, which measure weight increase (and caloric equivalent and chemical composition) of net production, (b) enclosure studies, involving measurements of CO_2 exchange in plastic enclosures of parts of ecosystems and (c) flux techniques based on measurements of CO_2 levels in the forest environment (Woodwell and Whittaker, 1968). The last method, which may yield extremely high quality data, requires the development of new techniques.

Cost estimates. The costs for source monitoring and research are given here with the annual operating costs estimated to be the number of scientific and technical staff multiplied by an assumed total cost per staff of $60,000. The capital costs are for data-processing equipment and laboratory test equipment:

	Capital Investment	Annual Operating Costs
	(thousands of dollars)	
Monitoring	40	120 (2)
Research	40	120 (2)
Total	80	240 (4)

The (   ) indicate the number of technical staff. The total cost would be approximately $300,000 for the first year.

THE PRINCIPAL ROUTES OF CO_2. The major route by which CO_2 is transported from one region of the earth to another is the atmosphere. Oceanic transport is much less effective.

The task of extrapolating global atmospheric CO_2 content calls for the institution of monitoring and research programs. The aim of the research would be to develop a quantitative understanding of CO_2 transport in the atmosphere as a basis for the extrapolation. The inputs to this transport model are the source data, whose characteristics in the future would be obtained by estimating future population increases and the use by this population of fossil fuel and wood.

The monitoring stations should be on land and in places where the measurements will be free of perturbations from local sources, wherever possible. If the research effort indicates a multi-source accumulation which may lead to regional climatic effects having possible widespread implications, then monitoring will also be necessary in these regions.

The research stations could be on land or in airplanes. Satellites seem to have potential here because they can, in principle, give continuous globally distributed data, including source effects, which the other monitoring systems cannot do. The satellite method is listed under the research effort because increases in precision and resolution are needed in the remote instrumentation which would have to be used; instrumentation for ground stations (Pales and Keeling, 1965) and aircraft is already developed.

Data-gathering requirements.

1.   Spatial resolution. For near term monitoring, we need to measure the total CO_2 content in a thoroughly mixed atmosphere, free of sources. Since the effects of the sources are not well understood, prudence suggests more than one station. Four widely separated stations are suggested, two in the Northern Hemisphere (Alaska and Hawaii) and two in the Southern Hemisphere (South Indian Ocean island and Antarctica). Such diversity in position tends to minimize perturbations due to some special regional effect.

From a costing view, this arrangement may be less satisfactory than a single land station plus one airplane. The latter could be used to minimize perturbations in "clean air" data due to vertically distributed sources. In fact, this is already being done by Swedish experimenters who use commercial flights mainly over the Arctic Sea (Bolin and Bischof, 1970). They have collected about 700 good samples from 50 flights. However, the data quality obtained by using airplanes seems to be somewhat inferior at present to that obtained from ground stations. In the short term, the latter approach is preferred, although the feasibility of using airplanes should be investigated, along with the use of statistical techniques to optimize data analysis.

For research, 12 stations (which include the 4 monitoring stations) have been suggested, of which 3 would be separated widely in longitude (Indian Ocean, North Atlantic, and South Atlantic) while the other 9 would stretch in the Pacific from the North to the South Pole with a spacing of 20° or 2,400 km. These stations would record effects due to atmospheric and ocean currents and possible source effects. This recommendation is similar to one made by the Commission for Air Chemistry and Radioactivity: "To monitor seasonal and regional variations adequately, the minimum number of stations for CO_2 and constituents of similar character is about 10." (Commission for Air Chemistry and Radioactivity, 1970.) The altitude dependence, to be obtained by using an airplane with improved instrumentation, is a research effort; the suggested vertical spatial resolution is about 2 km with several horizontal passes at different altitudes taken during the flight. This kind of research data coverage would be quite expensive.

2.   Temporal resolution. Data have been taken almost continuously for some years at a station on Mauna Loa (Pales and Keeling, 1965). In further monitoring it might be possible to collect data at greater intervals, leading to decreased data-processing costs, by using appropriate statistical techniques.

3.   Accuracy. Data taken to date at Mauna Loa show diurnal variations of ±0.5 ppm, seasonal variations of ±3.5 ppm, and a mean annual increase of 0.7 ppm (Pales and Keeling, 1965). Present indications are that changes of order 0.1 ppm in the annual rate of increase are significant in the extrapolation problem. The essential requirement is that there should be no systematic drift in reference standards, or systematically changing errors in operation of the instrument. An accuracy in individual readings of ±0.1 ppm is sought at the Mauna Loa station. It seems advisable to continue this practice at all monitoring stations for the present. In time, some relaxation, perhaps to ±1 ppm, might be permissible.

Data-gathering techniques.

1.   Monitoring. The *in situ* techniques are ideal for monitoring in the immediate future. They are in general extremely accurate and very economical in a ground station. Occasionally, impurities

in the intake pipes will degrade the data. Also "flask" data are about 1 ppm higher than continuous intake data for reasons which are not understood. The air sample is cooled to separate water vapor and the CO_2 concentration is measured by infrared techniques.

2. Research. When global coverage is desired, then *in situ* measurements become impractical and remote measurements must be made. Such measurements become very difficult when an accuracy of ± 0.1 ppm or even ± 1 ppm is desired. This requires that research be done on the most appropriate techniques for making the measurement: passive (no radiation source) or active (use of a radiation source); use of visible, near infrared, or far infrared wavelength bands; vertical viewing or horizontal viewing; and so forth. These, and other appropriate techniques should be placed in the study phase category. As noted earlier, this would involve feasibility studies which would include some instrumentation testing on the ground and in an airplane. These techniques will most likely be useful in obtaining data on other gaseous pollutants and particles.

Cost estimates. A suggested cost schedule for the "route effort" in the first year is as follows:

	Capital Investment	Annual Operating Costs
	(thousands of dollars)	
Monitoring*	2,000	960 (16)
Research†	600	480 (8)
Total	2,600	1,440 (24)

the * indicates that the Mauna Loa station (upgraded at a cost of about $150,000) and 3 new stations (each with an initial capital investment of about $600,000) would be prepared for monitoring during the first year; † indicates that the first year's research effort is divided into two parts. The first is a site selection study for the other 8 sites. This effort would use about 3 staff and $100,000 (for site selection experimental equipment) in capital investment the first year. The second effort is the beginning of an experimental program to use advanced technology in instrumentation and station platforms, which would involve approximately $600,000 for instrumentation and 5 staff; ( ) indicates the number of technical staff. The total cost of the CO_2 atmospheric route program would be approximately $4 million in the first year.

THE PRINCIPAL RESERVOIRS OF CO_2

Data-gathering requirements and techniques. Since the ocean is the principal reservoir, a high priority is placed in gathering data pertaining to its characteristics. The ocean's ability to store CO_2 varies very significantly with surface layer temperature. Furthermore, an exchange of surface water with deep ocean water, which is rich in CO_2, will cause the ocean to release CO_2 to the atmosphere.

The latter upwelling phenomenon occurs in three areas: Southwest of Greenland, the Norwegian Sea, and the Weddell Sea. These areas should be surveyed annually by oceanographic ships that would look for intermittent large-scale upwelling of the deep waters: the required measurements would be of C^{14}, the partial pressure of CO_2, temperature, and salinity.

The average temperature of the top 400 meters of the ocean should be monitored with sufficient accuracy to observe $0.5°C$ changes from year to year. Sea surface temperature is intensively measured for meteorological purposes (8,000 observations per day in the Northern Hemisphere shipping lanes) and there are plans in the World Weather Watch program for use of satellite methods and automatic data buoys. Meteorological requirements for surface water temperature measurements supply data sufficient for the needs of CO_2 monitoring.

Available data on subsurface temperatures are less satisfactory, but there are several deepwater stations already in use. These are ships that use submerged thermographs (thermistors) to measure temperature to $\pm0.3°C$ down to levels of 300 meters, with 0.3-meter vertical spatial resolution. A study of the configuration of observing methods best suited to meet all requirements for ocean water temperatures may soon be needed.

The major uncertainty in the process of extrapolating atmospheric CO_2 content is allocation between the ocean and the biosphere of that proportion of the source which does not remain in the atmosphere. This proportion will almost certainly vary over time as fossil fuel is burned. At present, it is uncertain within a factor of 2. This uncertainty could be decreased by measuring the rate of downward mixing of ocean waters from the surface to depths of about 600 meters or more. This might be obtained by following the C^{14} released by the past nuclear tests. (The present C^{14} levels are 150 to 200 percent of the pretest values.) The necessary data could be obtained by 3 meridional transects, one in each of the three major oceans, in which C^{14} could be measured at different latitudes to depths of about 1,000 meters. The data should be taken with a 3-year spacing at first, then 5 years, then every 10 years. The first year would involve only the transect in the Pacific.

Cost estimates. A suggested approximate cost schedule for the ocean reservoir effort in the first year follows:

Required Information	Capital Investment	Annual Operating Costs
	(thousands of dollars)	
Upwelling	150	220 (6)
Temperature	0	120 (2)
C^{14}	100	620 (2)
Total	250	960 (6)

In this schedule the temperature measurement effort is a study only and involves no capital investment. The upwelling and C^{14} measurement efforts involve ship costs that were estimated at the normal rate of $5,000/day. The capital investment for these two efforts includes the costs of instrumentation and data processing. The cost for the ocean program in the first year is approximately $1.2 million.

The total approximate cost of the CO_2 monitoring effort, based on these figures, is $5.5 million for the first year. It must be emphasized that this figure is to be used as a guide only, and is not based on a thorough cost analysis. This cost does not involve coupling to the other efforts in monitoring (which, of course, must be done) and is not realistic from that point of view.

Recommendations

1.   We recommend continuing attention to improving our estimates of present and future consumption of fossil fuel.

2.   We recommend similar attention to changes in the mass of living matter and its decay products.

3.   We recommend continuous measurement and study of the carbon dioxide content of the atmosphere in a few stations remote from sources—specifically 4 stations—and on some airplane flights. We particularly recommend that the existing record at Mauna Loa Observatory be continued indefinitely.

4.   We recommend continuing systematic study of the partition of carbon dioxide between the atmosphere, the oceans, and the biomass. Such research might require up to 12 stations.

The Particulate Load of the Atmosphere

The emission of solid and liquid particles into urban atmospheres and the formation of particles there by chemical reactions between emitted gases yield what is to many people the most obvious and obnoxious sign of "atmospheric pollution"—smog (National Air Pollution Control Administration, 1969; Cadle, 1970). Emissions are extensively monitored on a local basis in many parts of the world. We are concerned here with the truly global problem: "What changes in the particulate loading of the atmosphere affect the world's climate?" (Mitchell, 1969; Shapley, 1953). The question as it affects monitoring becomes, "Is the particle content of the atmosphere changing in a systematic way as a result of man's activities?" Monitoring can, at least in this instance, also help considerably with the further questions: "How serious is the climatic threat? How can we best correct the trend?"

Globally the particle load affects climate by changing the radiation balance of the earth. It acts in two ways: by changing the amount of solar energy available to the earth, that is, by changing the earth's albedo (reflectivity); and by changing the internal redistribution of the solar energy (CO_2 acts only in the latter way). Atmospheric particles have readily observed optical effects, which we monitor by optical devices, but these methods do not give us the opportunity to identify the sources of the particles. Chemical methods do permit identification and thus provide the data for rational discussion of abatement. A very large proportion of atmospheric particles, both natural and man-made, are not emitted as particles but are formed within the atmosphere by chemical reaction between trace gases. Hence, monitoring the particles implies monitoring these trace gases (Cadle, 1966, 1970).

MONITORING PARTICLES BY OPTICAL METHODS. We are concerned with measurements of solar radiation or of similar radiation from artificial sources in the atmosphere, and we wish to isolate the effect of particles on this radiation. Particles are transported by the atmosphere, and their reservoir is the atmosphere. The residence times of particles are hours to days in the troposphere, weeks to years in the stratosphere. Because of the short tropospheric residence times, spatial variation of the total particulate load of the atmosphere is very high. Cities and windy deserts are major sources, and the load is high in their vicinity. The global monitoring problem can be attacked by choosing a few sites remote from surface sources, by making simple measurements at many sites and examining the results statistically, or by standing off in space and looking at the earth as a whole.

The last of the three methods is the most fundamentally sound attack. The potential climatic effect of warming or cooling would be the result of a change in the earth's albedo (McCormick and Ludwig, 1967). The proposal is to measure this change to the required accuracy by satellite. The difficulty is that estimates of the change suggest it is likely to be small and to build up slowly. So, although we believe that satellite measurement of an annual average of the whole earth albedo with a precision of a fraction of 1 percent may just be feasible, we suspect that an attempt to establish a trend might meet the problem of a high noise-to-signal ratio solvable only by integration over decades.

The first of the three methods, the few remote sites, involves the measurement of solar radiation at the earth's surface, to include both the radiation received directly from the sun and the diffuse radiation scattered by the atmosphere. Analysis of records on cloudless days is required. It is possible to deduce approximately the overall scattering and absorptive properties of the total atmospheric load of particles (Robinson, 1962). Comprehensive solar radiation monitoring equipment, such as that which would be required, has been obtained for the Mauna Loa Observatory at a cost of $50,000. We recommend use of about 10 units of this type.

Trials have shown the potential usefulness of the lidar (optical radar) technique, which displays the intensity of the scattered light returned from the atmosphere against the height of reflecting particles. Experiments are required to establish the long-term stability and intersite comparability of these methods. In principle, they are a valuable supplement to solar radiation records at remote sites (Collis, 1966). Equipment costs for one site would be on the order of $100,000.

The method of the widespread collection of observations of low precision relies on the absence of *systematic* error. Continued attention to centralized standardization of the measurements is required to avoid such error. A monitoring network is now in operation, using the Volz photometer, which measures the intensity of direct sunlight at wavelengths defined by narrow-band interference filters (Flowers, McCormick, and Kurtis, 1969). The readings are readily taken by comparatively unskilled observers and do not rely on the existence of completely cloudless skies. The cost of the instrument itself is on the order of $100, and the time consumed is only a few minutes per day. Comparison with centrally maintained standards approximately twice a year is essential.

A global monitoring system should ideally involve all the above-mentioned types of measurement. Widespread use of the Volz photometer would produce an indication of any trends in major sources of particles. Solar radiation measurements would introduce a higher degree of precision in estimating any global trends and would allow some discrimination between scattering and absorption—important in estimating the effects on global albedo. Lidar measurements isolate the effects in the important lower stratosphere region. The feasibility of long-term precision monitoring by satellite of global albedo should be examined.

The estimated costs for integration with existing meteorological and other networks, with staff costs estimated on time-sharing basis are as follows:

500 Volz Photometer Sites

Equipment	$75,000
Annual running cost	$75,000

10 Solar Radiation Sites

Equipment	$400,000
Annual running cost	$250,000

4 Lidar Sites

Equipment	$400,000
Annual running cost	$250,000

Feasibility study of Albedo Satellite (no hardware) $100,000

MONITORING PARTICLES AND GASEOUS PRECURSORS OF PARTICLES BY CHEMICAL METHODS

Sulfate and nitrate particles. The organic particles, sulfates, and nitrates are major constituents of atmospheric aerosols of which a proportion is man-made (Junge, 1963; Cadle, 1966, 1970; Robinson and Robbins, 1970). They may affect the climate through their contribution to turbidity, their absorption of radiation, and their activity as condensation nuclei. The human sources of sulfates and nitrates are the combustion of sulfur-containing fuels and nitrogen fixation during combustion. The former produces sulfur dioxide, which is oxidized and hydrated in air to form sulfuric acid droplets. The latter produces nitric oxide (NO), which is oxidized and hydrated to form HNH_3 and nitrates (Cadle and Allen, 1970). There are numerous natural sources of sulfur dioxide and sulfates, such as volcanoes, the oceans, and forest fires. Forest fires and lightning are natural sources of oxides of nitrogen. Organic particles are formed by chemical reactions in smog and are emitted directly by various combustion processes (natural and man-made).

Probably the best way of monitoring sulfates and nitrates, at present, is at surface monitoring stations, collecting particles on polystyrene fiber filters (Cadle and Thurman, 1960). The filters are extracted and the extracts analyzed using colorimetric techniques. The monitoring for organic particles could use glass fiber filtration followed by extraction with benzene, evaporating the benzene, and weighing the residue. Total particulate concentrations are obtained by weighing the filters before and after sampling. Since these substances are atmospheric particles, their residence time in the atmosphere is relatively short and a large sampling network is advisable, perhaps 100 stations. Ten percent accuracy is satisfactory. One sample should be taken at each station every 2 weeks. The costs are itemized below:

Hydrocarbons. Certain types of hydrocarbons react with ozone, or, in the presence of sunlight, with oxygen and nitrogen oxides, to form particles (Cadle, 1966, 1970; Cadle and Allen, 1970).

Hydrocarbons are emitted by both urban and biosphere sources. Methane is not of concern because it is not a source of aerosols; terpenes from forested areas can be oxidized to form aerosols (Cadle, 1966). Only a very small fraction of the hydrocarbons from urban pollution have structures such that they form organic aerosols in significant amounts. Total emission of these olefins probably is only a few percent of the global terpene emissions.

We know very little about the half-life of hydrocarbons in the atmosphere. The reactive ones may have much shorter lifetimes than the others, so perhaps 100 stations are required. One sample should be collected every 2 weeks. Accuracies of ±20 percent may be obtainable if a concentration technique is workable. The most practical monitoring technique would be to sample those higher molecular weight hydrocarbons of interest on a solid substrate such as silica gel and transport these samples back to a central laboratory for analysis on laboratory gas chromatographs equipped with flame ionization detectors. The capi-

	Each	Total
	(thousands of dollars)	
1. Cost of Central Laboratory	$400*	$400*
2. Laboratory for manufacturing polystyrene filters	100	100
3. Sampling equipment	2	200
4. Cost (annual) of operating stations	50	5,000

* This assumes that many other substances would be monitored at the stations and the cost is for all monitoring.

tal investment per station specifically for hydrocarbons will probably be about $2,500. Some direct aircraft and balloon sampling can also be undertaken.

Nitric oxide and nitrogen dioxide. These oxides are converted rather slowly to particulate nitrates, but probably more important to particulate formation is the role oxides play in the photochemical formation of particles from hydrocarbons. Urban air pollution contributes appreciable amounts of nitrogen oxides. The rates of conversion to particulate nitrate appear to be slow, so conversions occur at least on a regional scale. It has been estimated that the biosphere over land areas contributes about an order of magnitude more nitrogen oxides to the global atmosphere than does urban pollution (Robinson and Robbins, 1970). However, the biosphere values are estimates based on only a few experimental measurements.

Spatial resolution should take into consideration the need to estimate biosphere contributions as well as upper atmosphere mechanisms of conversion of nitrogen oxides to nitrates. One sample every 1 or 2 weeks should be sufficient at each of 100 ground level monitoring sites. Aircraft sampling for nitrogen oxides also should be included in an overall aircraft sampling program. Accuracies of about ±20 percent would appear sufficient for initial monitoring efforts.

Nitrogen dioxide can be determined by colorimetric methods: these methods are suitable for sampling periods up to 2 hours (Intersociety Committee, 1969). Nitric oxide is analyzed by the same procedure as nitrogen dioxide after oxidation by a chemical substrate. The same sampling schedule should be used for nitric oxide as for nitrogen dioxide. The only now-available long-path or remote correlation instrument for nitrogen dioxide requires evaluation concurrently with the colorimetric procedures at nonurban sites.

The capital investment specifically for NO and NO_2 would be about $2,500 per station for about 100 stations.

Hydrogen sulfide and sulfur dioxide. Hydrogen sulfide is readily oxidized to sulfur dioxide, which in turn is converted to sulfuric acid aerosol or particulate sulfate (Cadle and Allen, 1970). These sulfur compounds in particulate form could be contributing an appreciable fraction of the increase in atmospheric turbidity reported in recent years.

Concentration levels of hydrogen sulfide on a global basis are very low, probably at or below 0.1 ppb (Junge, 1963). Sulfur dioxide occurs at a few tenths of 1 ppb on a global basis. These low levels also indicate effective conversion to sulfate on a global scale. Sulfur dioxide causes plant damage effects in many local and regional areas around the world.

Hydrogen sulfide is the major biosphere source of sulfur compounds. The conversion of hydrogen sulfide to sulfur dioxide over the land appears to be so rapid that the detection of any hydrogen sulfide over the oceans or polar regions is unlikely. Sulfur dioxide is produced both by oxidation of hydrogen sulfide and from urban pollution by combustion of sulfur containing fuels and smelting processes. The amounts of sulfur produced from urban sources in the United States will be extensively inventoried as part of the implementation of air pollution control programs.

A network of 100 stations is recommended and an accuracy of 20 percent is desirable if it can be achieved.

No chemical method exists that is acceptable for measuring very low levels of hydrogen sulfide. Sulfur dioxide can be measured by variations on the colorimetric Schiff reagent method referred to as the West-Gaeke procedure in urban air pollution measurements (West and Gaek, 1956). Collection of air samples in the liquid reagent over periods up to 24 hours can be conducted satisfactorily. Long-term sampling followed by automated analysis is also feasible. If the methods are not automated, the capital investment per field station will be about $1,000. This assumes the existence of the central laboratory mentioned earlier.

Satellite monitoring may someday be feasible. It will probably be several years before it is available, even if it can be made sufficiently sensitive.

Ammonia. Ammonia combines chemically with the sulfuric acid droplets resulting from sulfur dioxide oxidation and hydration to form ammonium sulfate particles. The nature of the particles in the atmosphere has a strong effect on the extent to which they scatter and absorb solar radiation and also their behavior as nuclei.

Both natural and man-operated sources of ammonia exist,

but there are many uncertainties. For example, it is not absolutely established whether the oceans are sources or sinks for ammonia (Junge, 1963). The soil can be both a source and a sink for ammonia, depending upon the pH. Alkaline soils tend to release ammonia and acid soils to absorb (react with) it. In general, natural ammonia results from the decomposition of amino acids by bacteria. Man-produced ammonia results from some combustion processes and from farming.

Ammonia is probably short-lived in the atmosphere (a matter of days), so, if possible, it should be monitored at about 100 stations every 2 weeks. Ten percent accuracy should be sufficient.

Simple, effective sampling and analysis techniques involving bubblers, reagents, and colorimeters (photometers) are available that can achieve the desired accuracy. Both sampling and analysis can be performed in the field at little additional cost over that for operating field stations for other trace constituents. As for most other trace atmospheric substances, satellite monitoring technology for ammonia has not been developed but might be preferable to ground stations if available.

Ozone. Ozone in the stratosphere is being monitored continuously and need not be considered here. However, tropospheric ozone is also of interest and is not being monitored. It reaches the lower troposphere from the stratosphere and also is produced by photochemical reactions involving the pollutants, oxides of nitrogen, and hydrocarbons. The latter is one reason for monitoring, since ozone is an indicator of photochemically active pollutants. Also, it reacts chemically with many hydrocarbons, both natural and man-made, to form particles (Cadle, 1966).

Since its half-life in the troposphere is short (hours or days), it should be monitored if possible at 100 stations. Commercially available units for continuous analysis cost a few hundred to a few thousand dollars per unit.

Recommendations

1.   We recommend extending and improving the precision of measurements of the transmission of solar radiation which are already being made (Volz photometer; total and diffuse solar radiation at the surface).

2.   We recommend beginning measurements by lidar (optical

radar) methods of the vertical distribution of particles in the atmosphere.

3.   We recommend studying the scientific and economic feasibility of initiating satellite measurements of the albedo of the whole earth capable of detecting trends of the order of fractions of 1 percent per 10 years.

4.   We recommend beginning a continuing survey, with ground and aircraft sampling, of the atmosphere's content of particles and those trace gases that form particles by chemical reactions in the atmosphere. This will require about 10 fixed stations for relatively long-lived atmospheric constituents, and about 100 fixed stations for short-lived constituents.

5.   We recommend about 100 measurement sites for monitoring each of the specific particles and gases by chemical means—including sulfate and nitrate particles, gaseous hydrocarbons, nitric oxide and nitrogen dioxide, hydrogen sulfide and sulfur dioxide, ammonia, and ozone. In general, one monitoring station could measure all particles and gases. However, a total of somewhat more than 100 stations would probably be necessary, since the combination of short residence time and some isolated sources which should be monitored would require stations that monitored only one or a few of the constituents.

Modification of the Composition of the Lower Stratosphere

The lower stratosphere is a quiescent region of the atmosphere. Residence time of gases there is on the order of a year. The purging effect of precipitation is also absent. Material formed in or injected into the stratosphere may therefore accumulate in higher concentrations than elsewhere in the atmosphere (Commission for Air Chemistry and Radioactivity, 1970; Martell, 1970).

It has been estimated that by 1985–1990, 500 supersonic transport (SST) aircraft will be operating, and computation shows that they will change the steady-state water content of the stratosphere by up to 10 percent (Machta, 1970). Higher transient local concentrations may lead to the formation of clouds. The aircraft will also produce particles, directly as carbon soot and indirectly following reactions between emitted sulfur, nitrogen oxides, and hydrocarbons, and the ambient ozone. Carbon monoxide is believed to be destroyed in the lower stratosphere by oxidation to

CO_2 in a complex photochemical reaction involving H_2O. Sulfur dioxide of tropospheric and man-made origin may reach the stratosphere and be oxidized to sulfate aerosol (Cadle et al., 1969; Cadle et al., 1970).

The emissions from the supersonic transport represent a relatively massive environmental change. Here is one instance where we can anticipate a potential unwanted side effect of technological advance. We feel it would be unwise not to institute monitoring now, before the change begins. Monitoring is required in the vicinity of projected major air routes, where the greatest concentration of contaminants is expected, as well as in regions remote from such routes.

WATER VAPOR IN THE STRATOSPHERE. If, as estimated, in the 1985–1990 time period, 500 SSTs fly 7 hours per day in the stratosphere, and each engine produces about 41,400 lb/hr of water vapor, the SSTs will increase the 3 ppm concentration of H_2O in the stratosphere to 3.2 ppm H_2O (see report of Work Group 1). This may have an effect on world temperature.

Several monitoring techniques based on direct sensing are at present available. The instrumentation is borne by balloons, rockets, or airplanes. Airplanes offer the best opportunity for systematic monitoring (Cadle et al., 1969; Cadle et al., 1970), but few types have the required performance in altitude and endurance. The RB57 aircraft of the U.S. Air Weather Service is suitable. Several types of instrumentation might be used. Mastenbrook has used balloonborne frost-point recorders, while instruments used on aircraft include frost-point hygrometers and others based on changes in the electrical properties of a surface as a result of the adsorption of water vapor.

Eventually, satellites may be able to do this monitoring. The needed technology is under development.

The costs for monitoring by aircraft are difficult to estimate, but if flights are available anyway, as at present, the capital investment could be as little as $10,000 for instrumenting two aircraft. The instrumentation would need to be carefully calibrated, and the calibrations repeatedly checked. Thus, instrument maintenance and analysis might cost $20,000 per year. Obviously, if special flights had to be financed, the costs would increase considerably.

Until trends are detected, 4 flights per year in the Northern and Southern Hemispheres should suffice. Since the instruments are continuous-reading devices, a given flight could cross both well-traveled and little-traveled regions. The accuracy should probably be about 0.2 ppm.

STRATOSPHERIC PARTICULATE MATTER. Particulate matter can be monitored on the same aircraft flights used for water vapor. However, the sampling will then be by filters, rather than continuous. Measurement accuracy will probably have to be ± 1 ppb by weight.

It may be appropriate to use two types of filters on each flight. Glass-fiber filters can be extracted with organic solvents following a flight to furnish concentrations of organic particles (largely from the SSTs), while the extremely high purity polystyrene fiber filters of the type now being prepared for such purposes at the National Center for Atmospheric Research may be used for sampling for $SO_4^=$, Si, and NO_3^- (Cadle, 1960). The Si measurements are included to indicate the amounts of soil mineral matter present. By weighing filters before and after flights, total particle loadings can be obtained.

Remote sensing by lidar (optical radar) can also be done, and probably ground-based lidar stations should be used to supplement the aircraft flights (see earlier in this section).

We are advised that SST operation may lead to the production of stratospheric clouds in heavily traveled areas in subarctic regions (see report of Work Group 1, section 1.2.4). Lidar appears to be the most promising device for monitoring this effect. While we estimate that 4 lidar sites would be adequate to supplement aircraft observations in the period prior to operational introduction of the SST, we anticipate the need for increased use of lidar monitoring as traffic density increases. In its present state of development lidar does not provide an absolute measure of particle concentration. Study of the lidar monitoring technique is recommended. As the detailed properties and quantities of the particles expected to result from SST operation are classified by theoretical and field study, the feasibility of their detection and monitoring by ground- and satellite-based remote sensing should be reviewed continuously.

The main capital investment in the initial phase will be for

(1) the clean rooms and other equipment for manufacturing the polystyrene fiber filters ($100,000); (2) a laboratory for conducting the analyses ($400,000*); and (3) the 4 lidars ($400,000). The annual operation excluding aircraft flights would cost on the order of $50,000 for analysis and $200,000 for the lidar sites.

CARBON MONOXIDE. Until recently there was no explanation of the apparently steady global atmospheric content of CO in spite of the contribution from vehicular emissions (Robinson and Robbins, 1969) and isotopic carbon measurements on carbon monoxide which could be interpreted to indicate that biosphere sources of carbon monoxide may produce much higher amounts of carbon monoxide than arise from urban pollution (Stevens, 1970). It is now thought that a removal process occurs in the stratosphere.

Carbon monoxide levels over unpolluted land and oceanic sites have been averaging in 1969–1970 0.1 to 0.2 ppm (Robinson and Robbins, 1969). Whether or not the steady-state concentration is increasing, determinations are needed of the effect of increased emission on the regions where CO is removed. If, as seems likely, a significant fraction of the carbon monoxide is being transported into the ozone layer of the stratosphere, the increase in the importance of the CO – OH reaction could be influencing the overall ozone reaction mechanism in the stratosphere (Hesstvedt, 1970).

Since carbon monoxide is long-lived (a few years) in the troposphere, monitoring is recommended at about 10 remote stations. Aircraft sampling could assist in supplementing this network by low-level sampling. Time resolution of one sample over 2 weeks per site should be adequate. An accuracy of at least ±5 to ±10 percent is essential to detect trends. No suitable wet chemical analytical methods are available. A highly sensitive and specific gas chromatographic technique is available for immediate use. A modification of the mercury-vapor-type analyzer has been used for carbon monoxide analysis in unpolluted areas. No instrument exists at present for remote measurements of carbon monoxide. The capital investment, other than that already in-

---

* This assumes that the laboratory will also be used to analyze samples collected at 10 to 100 ground-based stations, for servicing instruments, and for data processing.

cluded in the central laboratory, would be $10,000 at each field site if gas-chromatographic techniques or the mercury-vapor-type analyzer were used.

Recommendations

We recommend early implementation of a special monitoring program to comprise

1.   Measurement by aircraft and balloon of the water vapor content of the lower stratosphere. The height range of particular interest is 55,000 to 70,000 feet. The area coverage required is global, but with special emphasis on areas where it is proposed that the SST should fly.

2.   Sampling by aircraft of stratospheric particles with subsequent physical and chemical analysis.

3.   Immediate development of techniques for monitoring trace amounts of gaseous hydrocarbons, sulfur dioxide, and oxides or nitrogen in the stratosphere.

4.   Monitoring by lidar of optical scattering in the lower stratosphere, again with emphasis on the region in which heavy traffic is planned.

5.   Monitoring of tropospheric carbon monoxide concentration.

3.4.3
Monitoring Oil

An extensive discussion of the problem of oil pollution of the world's waters is included in the report of Work Group 2.

Techniques for Monitoring

Oil pollution is found in the ocean in at least six forms that require specific monitoring methods:

1.   Heavy contamination from initial spills.

2.   Thin films derived from the spreading of thick petroleum films and from biological sources. These are from 100 to 1,000 angstroms thick and are too thin to detect by methods that depend on interference effects or on the action of the refractive properties on visible or infrared electromagnetic waves. However, there are numerous physical effects of such films on the water, of which the most familiar is the smoothing of capillary ripples (Adam, 1941). This smoothing is easily photographed, particularly in the sun's glitter. At vertical incidence the rippled areas

are difficult to distinguish from smooth areas. In this case, and at night, infrared detection is feasible and convenient.

3.   Dispersed oil where molecules are so thinly dispersed on the water as not to be in close molecular contact. This does not constitute a film, and such aggregates are virtually indetectible *in situ;* samples must be skimmed off for chemical or physical manipulation.

4.   Lumps of oil some tens of millimeters across are formed when a thick film is broken up chemically and physically on the sea. These lumps appear to have a residence time on the sea surface of several months. They are probably consumed by biological and chemical reactions and eventually sink to the bottom.

5.   Oil is found in the volume of the sea or on the bottom, either in dissolved form or absorbed to particles or bubbles.

6.   Oil is found ingested in marine flora or fauna.

Of this list, the thick films are all too conspicuous and can be detected or monitored to any desired degree. They are of local distribution and occur most frequently near loading or drilling operations. Ship patrols are well suited to recovery of samples for chemical assay, while aircraft with or without air-to-sea sampling capability are convenient for rapid reconnaissance over large areas. For satellite monitoring, photographic or return-beam vidicon systems aimed at the sun's glitter path are practical, and photographic systems are already proved detectors of oil films.

The most obvious monitoring problem posed by thin films lies in the difficulty of discriminating ubiquitous thin, perhaps innocuous films, from the potentially dangerous thick ones. Methods that monitor in immediate contact with the surface or from low flight altitudes can discriminate between the two by observing interference effects (Newton rings, for example). This is a major advantage of these methods.

It does not at present appear appropriate to suggest a specific monitoring program for oil lumps on the sea (Horn, Teal, and Backus, 1970).

In the foregoing, techniques have been outlined that locate but do not usually identify the source of oil films. The latter requires physical capture of a sample for spectrographic, chromatographic, and other chemical analyses.

The methods now in use or under development for oil film identification consist of stripping it from the sea by using its adhesion to a wire screen or to a rotary drum, or of separating water from oil by centrifuging, or by scavenging as by ferric hydroxide sol. Other methods can doubtless be devised (Garrett, 1962, 1964; Harvey, 1966).

Recommendations

We do not make specific recommendations for widespread monitoring of oil. However, we do recommend more intensive investigations of local spills and effects. We think that any provision for global monitoring should await the outcome of base-line studies, such as the one proposed previously in this report, and of further investigation into the relation between oil and the transport and concentration of chlorinated hydrocarbons.

3.4.4
Monitoring Surface Changes
Potential Surface Changes

As the human population of the earth increases, greater and greater changes in the surface features of the planet take place. Forests are cleared for use as farmland, farmland is buried under the concrete of cities and roads. All the fossil and nuclear energy converted by man is ultimately dissipated as heat into the atmosphere and surface waters. These and other changes to the physical geography might ultimately affect the climate of the planet. Examples of land use or surface changes that, on a large enough scale, could have a global effect are the following:

1. Change of vegetation type. Changes from tropical forest to agricultural land may have a significant effect on the ability of the biosphere to absorb part of the carbon dioxide released to the atmosphere as a result of the burning of fossil fuel. It is also possible that the forests play a significant role in the transfer of water from the surface to the atmosphere in the tropics. Large-scale changes in the amount of forestland in these regions might have a significant effect on the general circulation of the atmosphere (see report of Work Group 1).

2. The spreading of deserts. New deserts may develop as a result of the diversion of groundwater, or existing deserts may change size as a result of man's activities on or near the deserts.

Changes in desert size might affect the climate in two possible ways: (1) through the change in reflectivity of the region, and (2) through the greatly increased amounts of dust introduced into the atmosphere by the winds.

3.   The filling of bays and estuaries. The tidal marshes that border bays and estuaries are extremely important to the growth of many marine species. For example, species as diverse as the shrimp or bluefish spend a critical part of their life cycles in these regions. As the marshes are filled to provide space for housing and other activities, the species that are dependent on these areas are lost, leading to the loss of important ocean fisheries. The trend should be monitored before a multitude of local encroachments has produced an irreversible global effect.

4.   Release of stored energy. This adds to the solar energy that drives the atmospheric circulation. In certain circumstances, it might lead to global as well as local climatic change. Much of the heat might be injected into rivers or estuaries, with serious effects on the life-forms existing there.

Monitoring Techniques

Land-use statistics can be derived from a multitude of sources, but over much of the world the data are probably inadequate. No system exists at present for compiling these data in a central file, for standardizing the data, and for filling in gaps in the data where they are known to exist (see report of Work Group 2). The major problems here seem to be data accuracy and obtaining data from the remote regions of the world.

A second approach to the monitoring of land-use changes is to utilize an airplane or an earth-orbiting satellite which can observe all or part of the earth. Satellite-type sensors, designed for land-use studies, are under development and are scheduled for flight during the next few years. By using multispectral scanning devices, instrument outputs that are compatible with large, high-speed digital computers can be obtained. The outputs can be interpreted not only as to whether the scene is soil, cropland, or forest but also as to type of soil, crop, or forest. Ground resolutions better than those required for climatological monitoring can be achieved now (Goldberg, 1969). Since these climatological surveys do not need the short time resolution that is required of these scanners for agricultural uses, it is quite unlikely that clouds would pose a significant problem. Surveys of 1-year duration should probably be flown at 5- or 10-year intervals. For this category the technology exists and will be in use in the near future. Costs for the instruments are of the order of $1 million (since the development costs have already been absorbed for the most part). It is quite likely that a survey could be conducted during the next few years using existing sensors and satellites. For use on a more limited geographical scale—for example, estuary studies— aircraft are of great utility.

Thermal Pollution

Thermal pollution might play a significant role in certain fisheries through its effect on the usefulness of the estuary as a spawning or nursery ground for fish that are found offshore in their mature state (Snider, 1968). Since thermal pollution is caused primarily by the discharge of heat from large generating plants, it can be monitored (at least in most countries) by monitoring the level of output and sources of electrical power. Estimates (possibly crude but quite meaningful) of the effect of a plant can be made from its power output and from knowledge of its method of heat dissipation. This information can often be ascertained simply from a knowledge of plant location. Monitoring thermal pollution, then, can be most easily done by a study of national statistics.

Recommendation

We recommend monitoring surface changes by attention to land use and economic statistics and by the study of the output of satellite observations of land use and heat balance.

3.4.5

Monitoring Nutrients in the World's Waters

It appears to us that the release of nutrients into rivers, lakes, and estuaries will contribute to a global problem if effects in coastal waters appreciably damage world fish populations (see report of Work Group 2). The problem is really comprised of a multitude of local problems, many of which are already being monitored locally. The techniques for such monitoring are sufficiently outlined elsewhere (National Academy of Sciences, 1969).

Recommendations

We recommend monitoring nutrients only on a detailed, local basis. Important quantitative information can be obtained by attention to production and use data. We recommend no global or coordinated actions. We emphasize that we recognize the current and potential importance of this problem and add our recommendation that its examination should be continued and expanded.

3.5

Considerations for Implementation

No unified action whose purpose is to change the course of man's impact upon the environment can be begun without some basis for an agreement that the change is necessary.

As we have reported on our ability to identify and measure global environmental problems, we have often referred to limitations that hinder an attempt to discover or present facts and conclusions that would provide that basis for agreement. We have said that sufficient base-line data do not exist; that existing mechanisms for keeping track of man's industrial activities do not include an "environmental effects" component; that we lack an adequate model of the workings of an environmental system; that we need better instrumentation; that more monitoring stations are needed.

Though we have said that such deficiencies should be remedied and have, in many cases, suggested specific steps toward a solution, we have generally not explored the organizational possibilities for overcoming the deficiencies. Nor have we attempted to survey the numerous organizations—governmental and non-governmental, national and international—which are already engaged in the monitoring, measurement, and research activities we have discussed.

Yet, it is our opinion that global problems are most efficiently studied and monitored on a global basis by a combination of national and international effort. We have viewed our work as the first stage of a larger exercise in exploring the dimensions, both scientific and political, of that effort. In what follows we recommend a general direction of inquiry for that exercise.

As we have said before in this report, we think that the most effective method for gathering information would be an integrated global network of fixed and mobile stations, each equipped to sample and analyze various pollutants in various subsystems of the environment. We also think that satellites can be important elements of such a network.

But we do not think that sufficient knowledge exists to establish such a system. A large amount of preliminary work must be done before any such attempt is made. We need comprehensive base-line surveys of the environment of the kind previously outlined in section 3.3 of this report. We need a comprehensive survey of current monitoring activities and facilities around the world. We face long discussions among all those involved in measurement and monitoring of the global environment on objectives, on what should be monitored, on standards for measurement and analysis of data, and on financial support. Assuming that any system which is set up will involve some combination of national and international activities, we also need agreements concerning divisions of responsibility and control within the system.

Such a list of requirements—one which is far from comprehensive—manifests the difficulties that stand in the way of any attempt at establishing a global network, difficulties that do not even begin to take into account the genuine differences in priority given to "the problem of the environment" by the various nations of the world.

Given the problems of negotiating, planning, and setting up such a network, it may be many years before it is providing decision makers with comprehensive, integrated information about the global environment. Because of this time lag, and because of the immediate need for new kinds of information about the problems which SCEP has identified, we stress that the setting up of this network should not impede continuous and necessary expansion of current measuring and monitoring activities.

Therefore, we recommend that an immediate study of global monitoring should be instituted to examine the scientific and political feasibility of integration and to set out steps for establishing an optimal system. One major component of that study should be a consideration of how existing and expanded monitoring activities, particularly those whose main value lies in continuity and homogeneity, can be merged into the integrated system with a minimum loss of data during the transition.

We believe that the proposed study might consider the following organization principles as it develops a structure for a monitoring system:

1.   As many national and international organizations as possible should be participating members of the system. Participants should be willing to modify their own monitoring activities in relation to the integrated monitoring plan.

2.   The participants should define the problems to be monitored and evaluated.

3.   The global monitoring system should be supervised by an agent of the participants—a truly international body, composed primarily of scientists and engineers, which would be responsible for determining how the measurements and monitoring should be carried out, but not with *what* problem to monitor.

   a. The supervisory body should not be separated either from scientific research or from the political scene; it should be able to give and take advice in both areas; its tasks and outputs must be very clearly defined. It should be obligated to maintain communications with the "user" of the monitoring information and to respond to "user" needs.

   b. The supervisory body should be initially charged with as great a responsibility for determining the process by which the monitoring system would be established as possible—the fewer constraints imposed on technical operations by the initial agreements setting up the system, the better.

   c. The supervisory body should have a technically competent monitoring staff to fill gaps and provide instruction.

4.   The global system should contain a center or centers concerned with the indefinite maintenance of physical, chemical, and biological standards and with the control of procedures.

5.   The global system should contain a "real time" data analysis mechanism. This would assure, for example, proper maintenance of measuring standards by allowing for prompt feedback to monitoring units in terms of modification of measurement parameters, levels of accuracy, and frequency of observation.

6.   The implementation of the global system by the supervisory body should proceed by

   a. The preparation of feasibility studies clarifying the objectives for measurement, examining the applicability of instrumentation, and completing a cost-benefit analysis of alternative approaches for solving the given problems.

   b. The preparation of comprehensive plans which would, among other things:

(1) Establish fiscal, hardware, and man-power require-ments.

(2) Clarify the nature of the problem, whether it was to monitor a well-defined environmental problem, assess the magnitude of a suspected problem, or provide infor-mation needed to understand the global environment.

(3) Define standards and procedures for sampling data processing and analysis.

(4) Define the output of the system.

Recommendations

1.   We recommend an immediate study of global monitoring to examine the scientific and political feasibility of integration of existing and planned monitoring programs and to set out steps necessary to establish an optimal system.

2.   We recommend the expansion of current measuring and monitoring activities in accordance with our recommendations in the rest of this report to satisfy the immediate need for new kinds of information about the problems SCEP has identified.

3.   We recommend that, because some components of what might ultimately be an integrated global monitoring system are so obviously needed, study and implementation of them be at-tempted independently of the investigation of an optimal global monitoring system. We specifically recommend a study of the pos-sibility of setting up international physical, chemical, and bio-logical measurement standards, to be administered through a monitoring standards center with a "real time" data analysis capability, allowing for prompt feedback to monitoring units in terms of such things as measurement parameters, levels of accu-racy, frequency of observations, and other factors.

References

Adam, N. K., 1941. *The Physics and Chemistry of Surfaces* (London: Oxford University Press).

Ayres, R. U., and Kneese, A. V., 1968. Environmental pollution, *Federal Pro-grams for the Development of Human Resources,* Vol. 2, a report submitted to the Subcommittee on Economic Progress of the Joint Economic Committee, U.S. Congress (Washington, D.C.: U.S. Government Printing Office).

Bolin, B., and Bischof, W., 1970. Variations in the carbon dioxide content of the atmosphere, *Tellus,* forthcoming.

Cadle, R. D., 1966. *Particles in the Atmosphere and Space* (New York: Reinhold).

Cadle, R. D., 1970. Atmospheric chemistry and aerosols, background paper prepared for SCEP (unpublished).

Cadle, R. D., and Allen, E. R., 1970. Atmospheric photochemistry, *Science, 167.*

Cadle, R. D., Bleck, R., Shedlovsky, J. P., Blifford, I. H., Rosinski, J., and Lazrus, A. L., 1969. Trace constituents in the vicinity of jet streams, *Journal of Applied Meteorology, 8.*

Cadle, R. D., Lazrus, A. L., Pollock, W. H., and Shedlovsky, J. P., 1970. The chemical composition of aerosol particles in the tropical stratosphere, *Proceedings of the American Meteorological Society Symposium on Tropical Meteorology* (unpublished).

Cadle, R. D., and Thuman, W. C., 1960. Filters from submicron-diameter organic fibers, *Industrial and Engineering Chemistry, 52.*

Callender, G. S., 1961. Temperature fluctuations and trends over the earth, *Quarterly Journal of the Royal Meteorological Society, 87.*

Carver, T. C., 1970. Estuarine monitoring program, *Report of the Subcommittee on Pesticides of the Cabinet Committee on the Environment* (Washington, D.C.: U.S. Government Printing Office).

Citron, R., 1970. *National and International Environmental Monitoring Programs* (Cambridge, Massachusetts: Smithsonian Institution), forthcoming.

Collis, R. T. H., 1966. Lidar: a new atmospheric probe, *Quarterly Journal of the Royal Meteorological Society, 92.*

Commission for Air Chemistry and Radioactivity of the International Association of Meteorology and Atmospheric Physics of the International Union of Geodesy and Geophysics, 1970. *Report to the World Meteorological Organization on Station Networks for Worldwide Pollutants.*

Feltz, H. R., 1969. *Monitoring Program for the Assessment of Pesticides in the Hydrologic Environment* (Washington, D.C.: Water Resources Division, U.S. Geological Survey).

Flowers, E. C., McCormick, R. A., and Kurtis, K. R., 1969. Atmospheric turbidity over the United States, 1961–66, *Journal of Applied Meteorology, 8.*

Garrett, W. D., 1962. Collection of slick-forming materials from the sea, Naval Research Laboratory Report 5761; also in *Limnology and Oceanography, 10.*

Garrett, W. D., 1964. The organic chemical composition of the ocean surface, Naval Research Laboratory Report 6201; also in *Deep Sea Research, 14.*

Goldberg, I., 1969. Design considerations for a multi-spectral scanner for ERTS, *Proceedings of the Purdue Centennial Year Symposium on Information Processing* (unpublished).

Harvey, G. W., 1966. Microlayer collection from the sea surface: a new method and initial results, *Limnology and Oceanography, 11.*

Hesstvedt, E., 1970. Vertical distribution of CO near the tropopause, *Nature, 225.*

Horn, M. H., Teal, J. M., and Backus, R. M., 1970. Petroleum lumps on the surface of the sea, *Science, 168.*

Intersociety Committee on Methods for Ambient Air Sampling and Analysis, 1969. Tentative method of analysis for nitrogen dioxide content of the atmosphere, *Health Laboratory Science, 6.*

Junge, C. E., 1963. *Air Chemistry and Radioactivity* (New York: Academic Press).

Keeling, C. D., 1970. Is carbon dioxide from fossil fuel changing man's environment?, *Proceedings of the American Philosophical Society, 114.*

McCormick, R. A., and Ludwig, J. H., 1967. Climate modification by atmospheric aerosols, *Science, 156.*

Machta, L., 1970. Stratospheric water vapor, background paper prepared for SCEP (unpublished).

Martell, E. A., 1970. Pollution of the upper atmosphere, background paper prepared for SCEP (unpublished).

Minzner, R. A., and Oberholtzer, J. D., 1970. Space applications instrumentation systems, National Aeronautics and Space Administration Technical Report C-136.

Mitchell, J. M., 1969. Climatic change—an inescapable consequence of our dynamic environment (paper delivered at the American Association for the Advancement of Science, Boston).

Murray, W. S., et al., 1970. National pesticide monitoring program, *Report of the Subcommittee on Pesticides of the Cabinet Committee on the Environment* (Washington, D.C.: U.S. Government Printing Office).

National Air Pollution Control Administration (NAPCA), 1969. *Air Quality Criteria for Particulate Matter* (Washington, D.C.: NAPCA).

Pales, J. C., and Keeling, C. D., 1965. The concentration of atmospheric carbon dioxide in Hawaii, *Journal of Geophysical Research, 70.*

Robinson, E., and Robbins, R. C., 1969. Sources, abundance, and rate of gaseous atmospheric pollutants (Menlo Park, California: Stanford Research Institute).

Robinson, E., and Robbins, R. C., 1970. Gaseous nitrogen compound pollutants from urban and natural sources, *Journal of the Air Pollution Control Association, 20.*

Robinson, G. D., 1962. Absorption of radiation by atmosphere aerosols, as revealed by measurements at the ground, *Archiv für Meteorologie, Geophysik, und Bioklimatologie, B.12.*

Rohlich, G., ed., 1969. *Eutrophication: Causes, Consequences, Correctives* (Washington, D.C.: National Academy of Sciences). See especially Chapter IV, Detection and measurement of eutrophication.

Sand, P. F., 1970. National pesticide monitoring program, *Report of the Subcommittee on Pesticides of the Cabinet Committee on the Environment* (Washington, D.C.: U.S. Government Printing Office).

Shapley, H., ed., 1953. *Climatic Change: Evidence, Causes, and Effects* (Cambridge, Massachusetts: Harvard University Press).

Snider, G. R., 1968. Nuclear power versus fisheries (paper presented at the annual meeting of the Isaac Walton League, Portland, Oregon).

Stevens, C. M., 1970. Natural and man-produced emissions of carbon monoxide (unpublished).

West, P. W., and Gaeke, G. C., 1956. Fixation of sulfur dioxide as disulfitomercurate (II) and subsequent colorimetric estimation, *Analytical Chemistry, 28.*

Woodwell, G. M., and Whittaker, R. H., 1968. Primary production in terrestrial ecosystems, *American Zoologist, 8.*

4.
Work Group on Implications of Change

Chairman
Milton Katz
HARVARD LAW SCHOOL

Robert U. Ayres
INTERNATIONAL RESEARCH AND TECHNOLOGY CORPORATION

John F. Brown, Jr.
GENERAL ELECTRIC RESEARCH AND DEVELOPMENT CENTER

Edward R. Corino
ESSO RESEARCH & ENGINEERING COMPANY

Willard A. Crandall
CONSOLIDATED EDISON COMPANY OF NEW YORK, INC.

John Franklin
GREENWICH, CONNECTICUT

Edward Hamilton
THE BROOKINGS INSTITUTION

Henry J. Kellerman
NATIONAL ACADEMY OF SCIENCES

George W. Rathjens
MASSACHUSETTS INSTITUTE OF TECHNOLOGY

Walter O. Spofford
RESOURCES FOR THE FUTURE

Howard J. Taubenfeld
SOUTHERN METHODIST UNIVERSITY SCHOOL OF LAW

Howard Wiedemann
DEPARTMENT OF STATE

Rapporteurs
Peter Katz
HARVARD LAW SCHOOL

Terry L. Schaich
FLETCHER SCHOOL OF LAW AND DIPLOMACY

4.1

Background and Perspective

When key pollutants have been identified, their effects ascertained, their sources found, their rates of accumulation calculated, their routes traced, and their final sinks or reservoirs located, the question of what to do about them will remain. In the context of current public opinion and social attitudes in the United States, a general disposition to eliminate or diminish the pollution may be taken for granted. An understanding of how to do so, what may be entailed in the effort, and what the consequences may be cannot be taken for granted.

"Residuals" or "waste" are generated in all stages of the production and consumption of goods or services. Residuals are a part of the process of production, in somewhat the same sense that the emission of CO_2 by a human being when he exhales is part of the process of breathing. "Residuals" become "pollutants" or an "environmental problem" of some kind and in some degree when they have harmful effects in the atmosphere, the oceans, or the terrestrial environment. "Harmful effects" are effects that are harmful to man, or to animals, plants, or inanimate objects or conditions that are important to man. Their importance to man may be biological, economic, religious, moral, aesthetic, or intellectual.

It must be recognized that such terms as "harmful" or "important" are charged with value judgments and are profoundly affected by the prevailing concept of man and his relation to the environment. This does not preclude objective analysis, but it throws light on what an objective analysis means in this context. An objective analysis in this sector of problems is one that at all times takes account of the fact that the meaning and weight of the terms of the analysis will vary with differences in value judgments and differences in the basic concept of man and his relation to the environment.

In essence, then, the environmental problem may be seen as a function of (1) the growth of population; (2) the development of technology and the application of technology in economic and social organization and activity; and (3) the priorities accorded respectively to the first-order effects of technology and the second-, third-, or higher-order effects of technology (side effects), as well as the care, skill, and imagination with which the potentialities of technology and of related economic and social organizations are developed and applied.

Although it is somewhat arbitrary to mark particular periods of history as the beginning or end of fluid, complex, and long-continued historical processes, it is customary to take the seventeenth century as the beginning of the scientific revolution and the eighteenth century as the beginning of the industrial or technological revolution. Both revolutions have been running continuously ever since, with periodic bursts of activity. The steeply accelerated rise in world population in the last two centuries is associated with, and has been made possible by, the rise and pervasive spread of science and technology. The ongoing scientific and industrial revolutions have spread from Europe and North America to other continents. The development or modernization of less-developed countries, which constitutes so important an aspect of the contemporary world, consists in major part of the absorption of the scientific and technological revolutions into those societies.

As populations have grown, and as the production of goods and services has multiplied through the expansion and increasing sophistication of technology and its application through new forms of economic and social organization, the emission of "residuals" or "pollutants" has correspondingly increased. In the earlier phases of the process, the population was preoccupied with the first-order effects. The American people wanted railroads, automobiles, electric power, fertilizer, pesticides, roads, central heating in homes, and air conditioning; they were comparatively indifferent to the concomitant increase in pollutants. In the main, they took the environmental effects in stride. To the extent that they thought or talked about it, they tended to regard the deterioration in the environment as the "price of progress." Their indifference was perhaps fortified by the comparative ease with which the propertied sections of the populace could escape from the pollution by moving to the better residential quarters of cities, to suburbs or countryside, and to the choicer beaches and mountain areas as vacation spots. There were important and discerning voices that deplored the seamy side of industrial development—reform and conservation efforts have a long history; protest exploded from time to time; and the legal system did recognize that certain applications of technology in particular places at particular times could be enjoined or subjected to the payment of damages as nuisances. But, in general, it may be said that the prevailing value system assigned an overwhelming priority to first-order effects.

In the current phase of the continuing technological revolution, a profound shift in values appears to be under way. Angry urban residents rally behind embattled mayors or civic organizations to check the construction of new highways, new jet airports, or new electric power plants. The Rachel Carsons and Ralph Naders receive widening attention and applause. A multiplication of bills and investigations in Congress relating to the environment and to the social implications of science and technology reveal a growing sensitivity among politicians to the new mood and the new possibilities. The shift in values brings with it a corresponding shift in the distribution of emphasis between first-order effects and side effects. The growing concern about the side effects—pollution—is soundly based on the facts of contemporary life, and the popular and political determination to give it effect may be expected to persist. But the need and the determination to bring the side effects under control do not import a loss of interest in the first-order effects. The actual and potential first-order accomplishments of technology remain necessary to support contemporary populations and retain a firm hold in the prevailing system of values. The shift in values, however, does entail a determination to achieve a better balance, and a better balance is needed.

A recent report by a Panel on Technology Assessment of the National Academy of Sciences stated the need in terms of "total systems effects." The report stressed a need to

. . . begin the task of altering present evaluation and decision-making processes so that private and public choices bearing on the ways in which technologies develop and fit into society reflect a greater sensitivity to the total systems effects of such choices on the human environment. How can we best increase the likelihood that such decisions . . . will be informed by more complete understanding of their secondary and tertiary consequences, and will be made on the basis of criteria that take such consequences into account in a timelier and more systematic way? And how can we do these things without denying ourselves the benefits that continuing technological progress has to offer, especially to the less favored portions of the human population? (Panel on Technology Assessment of the National Academy of Sciences, 1969).

4.2

Global Effects and Key Pollutants

As has been explained, this Study of Critical Environmental Problems (SCEP) has defined its mission in terms of key pollutants that have global effects. "Global effects" have been taken to compromise effects on climate and on ocean and terrestrial ecology, together with such effects as recur on a significant scale in many countries in a kind of worldwide pattern. The "key pollutants" are those whose global effects are such as to make it especially important to bring them under satisfactory control. Other reports of Work Groups in this book have assayed the present knowledge concerning residuals and their effects and have projected lines of inquiry to fill gaps in data, settle controversies, and resolve doubts. A number of residuals have emerged as prime subjects of attention. These include carbon dioxide; particulate matter; sulfur dioxide; oxides of nitrogen; toxic heavy metals (lead, mercury, arsenic, chromium, cadmium, nickel, manganese, copper, zinc); oil; chlorinated hydrocarbons, especially DDT and polychlorinated biphenyls; other hydrocarbons; radionuclides; heat; and nutrients.

This section focuses upon the problems, possibilities, and implications of remedial action. For its purpose it will assume that the character of the foregoing residuals as key pollutants with global effects either has been or in time may be established with a sufficient approximation of certainty or degree of probability to warrant remedial action. In exploring the implications of remedial change, we will seek to describe the kinds of questions that will arise and to suggest some approaches that may be useful. The questions will be explored first as they may arise within a single nation, illustratively the United States; then in an international context, where international collaboration may appear necessary or appropriate; and finally in the context of the relations among the advanced industrial societies and the less developed countries.

4.3

Implications of Change Within the United States

Remedial action may take the form of measures to eradicate the pollutants, to reduce their volume, to neutralize them by converting them into harmless residuals, or to shield people, animals, plants, or things from their effects. Remedial measures may be instituted at the sources of the pollutants, on the routes along which they spread, or in the reservoirs or sinks in which they finally accumulate. In our judgment remedial efforts should be concentrated at (1) the sources or the points in the processes of production, distribution, and consumption that generate the pollutants—for example, factories, power plants, stockyards, bus lines; (2) the protosources or earlier points in the sequence of processes that set the conditions leading to the emission of pollutants at a later stage—for example, the manufacturers of automobiles which emit pollutants when driven by purchasers, or the brewers of beer sold in nonreturnable cans that are tossed aside by the consumer; and (3) the secondary sources or points along the routes traveled by pollutants where they are concentrated, with or without treatment, before moving on to reservoirs—for example, sewage treatment plants or trash collection centers.

In the complex interrelationships of modern industrial society, the sources will often be multiple and the proportion of their respective contributions hard to ascertain. There will also be much room for differences of opinion as to which points in the processes of production, distribution, and consumption should be identified as sources or protosources and which points along the routes should be regarded as secondary sources. For practical purposes, however, informed judgment will be able to make workable identifications in most circumstances.

We believe it advisable to adopt a principle of presumptive responsibility for sources, protosources, and secondary sources. The principle is not put forward in any pejorative sense and does not connote any element of blame or censure, nor is it intended to foreclose judgment concerning where the economic costs of correction should ultimately be borne. The principle is practical in concept and purpose. It is intended as a point of departure for action and analysis. In part, it rests on the same basis that underlies familiar scientific practice and management practice: if something goes wrong, trace it to its origin and make the correction in terms of the cause. In part, it rests on the hypothesis that the sources, protosources, and secondary sources of pollution would typically be in the best position to take effective corrective measures, whether alone or with help and support from others. In part, it rests on the view that the nature of the remedial measures that may be available, the criteria for choice among them, the means by which they may be instituted, the economic costs of doing so, and the organizational or social adjustments that may be entailed can best be discerned and appraised at the sources, protosources, and secondary sources.

4.3.1
A Roster of Sources by Types

The classification of the sources, protosources, and secondary sources by principal types can serve as a convenient frame of reference. We suggest a classification that includes large industrial enterprises, small industrial enterprises, electric and gas utilities, transportation enterprises (railroads, airlines, bus lines, truck lines, taxi fleets), large farms organized as corporate enterprises, small farms, commercial enterprises (stores and service establishments), domestic activity (automobile driving by individuals and families, home heating, domestic waste). It should include local governments (with special emphasis on the collection and disposal of wastes through sewage systems and solid waste disposal methods), state governments (with special reference to their direct involvement in production and distribution such as highway construction, truck fleets, police cars, state liquor stores), and the national governments (with special reference to their activities, civil or military, involving the development or application of technology, such as defense, space exploration and utilization, development and control of nuclear energy, weather forecasting and weather modification, and development of transportation systems such as the supersonic transport).

Figure 4.1 may serve as a useful device to help keep the categories of sources and the list of key pollutants in mind, as well as to suggest their interrelationships.

4.3.2
Types of Remedial Change at the Sources and
Their Implications: An Overview

We believe it useful to outline the types of remedial change available for consideration in relation to the various sources and to sketch their implications in broad categories. Although such an overview cannot serve as a basis for action, it can be useful as a

Figure 4.1 Key Sources and Pollutants

Roster of Sources / Key Pollutants	CO₂	SO₂	NOₓ	Particulates	Chlorinated Hydrocarbon Biocides	Other Hydrocarbons	Toxic Metals	Oil	Radionuclides	Heat	Nutrients
Large Industrial Enterprises											
Small Industrial Enterprises											
Electric and Gas Utilities											
Transportation Enterprises											
Large Corporate-Type Farms											
Small Farms											
Commercial Enterprises											
Domestic Activity											
Local Governments											
State Governments											
National Governments											

general analytical model and checklist. For the purposes of such a model, it will be assumed that the source of a key pollutant has been located; that the effects of the pollutant warrant serious remedial action; that appropriate public institutions exist to initiate such actions; that there is a will to act in the appropriate public institutions and in the general community and that the source itself is prepared in principle to contemplate remedial measures. In circumstances so congenial to corrective action, what kinds of change should be taken into account in choosing among possible courses? They will vary depending on the nature of the source. In the case of industrial enterprises, for example, the possibilities may be listed as follows:

1. Changes in the input of raw materials.

2. Changes in the sources of energy used—for instance, fossil fuel or electric power (which may itself be derived from fossil fuel, waterpower, or nuclear power).

3. Changes in the processes of production, including

    a. Changes in organization or method.

    b. Recycling and the absorption of potential residuals in the manufacture of by-products.

    c. Changes in the treatment of residuals—capture, collection, storage, transportation, controlled disposal. (Changes of this type overlap with those in item b.)

4. Changes in the end product.

5. Changes in the quantity of the end product produced.

Since the changes will be instituted for reasons unrelated to the ordinary business calculations of profit and loss, the changes will usually involve economic costs, and the costs may be large in relation to the scale of the source enterprise (or protosource or secondary source, if the change is made at such a point). In broad social terms these costs will have to be weighed against the benefits to be anticipated from the curtailment of the pollutant. Such a cost-benefit analysis will be affected by the way in which the costs are allocated. If it should prove practicable to meet the costs out of gains to be derived from a by-product process, the costs to the enterprise will be compensated by gains for the enterprise. If the costs are absorbed by the source enterprise in the sense of being met out of what would otherwise be net income, or if they are passed on to the purchasers of the end product as may be expected whenever the competitive situation permits, the balance of cost against anticipated benefit typically will be asymmetrical, in that the social benefits will accrue to one sector of the society (all those who benefit from the elimination or reduction of the pollution) and the economic costs will be borne by another (the stockholders or customers of the source enterprise). The asymmetry will derive from a preexisting asymmetry caused by the pollution itself, but the nature and the distribution of the costs and benefits will be reversed. The social costs of pollution are borne by the sectors of society that suffer from the deterioration of the environment, while economic benefits are enjoyed by the source enterprise and the purchasers of its product.

In the terms of familiar economic analysis, the injuries done to the environment and to the society by pollution are "external costs" or "social costs," not taken into account in the ordinary business calculations of income and expense. They have been "external costs" not for reasons inherent in the nature of things or derived from the fundamentals of economics but because the legal system has so provided. In the large and in the long run the legal system reflects the prevailing values and priorities of the society, and it changes as social values change. By appropriate changes in the law, costs previously social and external can be internalized as economic costs to an enterprise that generates them. Such an internalization would represent one application of the priniciple of source (or protosource or secondary source) responsibility.

The principle as we propose it, however, is a principle of presumptive responsibility for sources, protosources, and secondary sources. Situations may arise in which the costs of a remedial change cannot be absorbed by the source (or protosource or secondary source) enterprise or paid out of gains from by-product operations, or passed on to customers. If the remedial change should nevertheless be compelled in such a situation, the source enterprise will fail. The termination of its activities could be viewed as one variant of a type of remedial change previously described as a "change in the quantity of the end product produced." Weighing all factors involved, such an outcome in some cases may be regarded as just and unavoidable. In other cases, however, apart from the resistance with which the source enterprise itself will obviously oppose such an outcome, it may be deemed unsatisfactory in larger social terms. The end product itself may be needed or desired by the society in undiminished quantity. In such an event it will be necessary to reappraise the balance between the cost of the change and the benefit anticipated from eradication of the pollutant. In the exercise of a prudent and well-informed social judgment, a decision might be reached to continue to bear the pollution. Alternatively, it might be found desirable to provide financial assistance to the source enterprise to effect the change—in short, to pay for the remedy by a subsidy out of public revenues.

The initial remedial change will entail consequences that may reach far beyond the economic costs to the source enterprise itself. If the type of raw material or energy previously used should be altered, the previous suppliers to the source enterprise will feel it. A remedial measure that curtails one pollutant may generate another no less detrimental, or residuals whose effects in the environment are unknown and must be ascertained. A shift in electric power production, for example, from generators fired by oil and coal to generators driven by atomic energy would eliminate emissions of carbon dioxide and sulfur dioxide but increase the discharge of waste heat into rivers or lakes and raise intricate problems relating to the disposal of radioactive residuals. As change spreads out from its point of initial impact, it can encounter and overturn deep-rooted social habits and patterns of behavior that have grown up around the usage of particular end products. The history of the industrial revolution has made us familiar with far-reaching readjustments imposed on sectors of the society by technological innovation seeking first-order effects. The manufacturers of horse-drawn buggies can bear witness to, and the decaying sections of our central cities stand out as reminders of, this experience. Comparable readjustments may be required by technological innovations motivated by a concern to diminish side effects.

We propose now to vary some of the simplifying assumptions we have made as a part of our general model: that the need for remedial measures affecting identified pollutants has been generally recognized; that a will to institute remedial measures is widespread in the society at large and shared by the relevant governmental institutions; that such institutions are appropriately designed and equipped for their tasks; and that the source enterprises themselves are prepared to acknowledge the need for remedial measures, at least in principle. These assumptions were intended only to facilitate the organization of our analysis in orderly stages and not to conceal vital aspects. In the actual processes of coming to grips with critical environmental problems, the source enterprises will often be found in the first instance to be unconvinced and indisposed to cooperate; public opinion may be confused and divided; the mood of the relevant governmental institutions may be uncertain; and the existing institutions may not be well designed to conceive and institute effective remedial measures. In such circumstances, how is understanding to be promoted and how are initiatives to be stimulated? By what means can reluctant participants be brought to take necessary or desirable steps? Who will do what?

For the purposes of our general overview it may be said that answers to such questions must be found, if they are to be found at all, within one or both of two vast and intricate processes. One consists of the interactions and mutual adjustments among innumerable separate points of decision, large or small, private or governmental, subsumed under the broad term "the market." The other consists of the interactions, adjustments, discussion, arguments, and varied measures of inducement and compulsion subsumed under the broad term "the political process." Both the market and the political process are profoundly affected by the nature and quantity of information available, the extent of its dissemination, and the manner in which the information is infused into the respective processes.

In this book we have presented much existing knowledge concerning key pollutants and their global effects and have identified important gaps in knowledge that can and should be filled. The existing knowledge has been collected by scientists and technical and professional personnel in universities, research institutions, industry, and governmental agencies. There are, however, several general needs for the development and use of knowledge:

1.   To foster a greatly extended and deepened search for knowl-

edge concerning critical environmental problems in both the private and the governmental sectors.

2. To improve the organization of such knowledge and to expand its dissemination to industry and agriculture; among scientific, technical, and professional groups; among national, state and local governmental institutions; and to the general public.

3. To bring increasing imagination and resourcefulness to bear upon the manner in which such knowledge may be articulated into the framework of market and political decision.

The first and second of the foregoing points speak for themselves. The third requires further elaboration. Its importance should not be underestimated. If the political process is at times unresponsive to scientific data, as it often seems to be and frequently may be in fact, the reasons may be varied. Among the reasons may be the form and style of the attempted input and inadequacies in the organization and staff of the legislative bodies to receive it.

Other sections of this Report reveal the gaps and uncertainties in existing knowledge concerning the volume and extent of pollutants, their sources, the routes along which they travel, the reservoirs in which they accumulate, and the scope and nature of their effects. We have already stressed the need to repair the deficiencies in information and analysis. It does not follow that remedial action must be postponed until scientific certainty can be achieved.

The political process rarely enjoys the privilege of certainty. It is accustomed to decisions in the face of some uncertainty made on the basis of a preponderance of evidence, or substantial probabilities, or a reasonable consensus of informed judgment either that a condition exists requiring remedial action or that risks of prospective harm are sufficient to warrant steps to preserve society's options. But it is wary of positive assertions that turn out to be unfounded and of experts who do not distinguish between their allegations within the field of their expertise and their allegations made as citizens commenting upon general matters. It will not assume that scientists and scholars are objective merely because they are scientists and scholars but will assess their objectivity in its own way and on its own terms.

Scientists, in presenting their reports or advice, will do well to separate out the "do know," the "don't know," the "can know," and their personal opinions. Personal opinions need not be avoided if they are clearly identified. An appearance of consensus should never be forced; but if a substantial consensus exists in fact among scientific peers, it can be acceptable and helpful if the inadequacies of the evidence on which it is based are fairly stated, the degree of dissent is identified, and dissenting opinions with their supporting analysis are incorporated in the Report. A pro-and-con format may be helpful. As far as practicable, scientific and technical data should be presented not only in the form appropriate for scientific and technical analysis but also in such a way as to relate the data to possibilities of corrective action.

In this connection, we also take note of proposals that have been made by the Subcommittee on Science, Research, and Development of the House Committee on Science and Astronautics, by the National Academy of Sciences, and by others to provide the Congress with additional and strengthened facilities for handling critical environmental problems and other problems rooted in technology and its implications. Comparable proposals have been made in relation to the legislatures of some states. While expressing no view concerning the comparative merit of particular proposed arrangements, we affirm the importance of providing legislative bodies with instrumentalities and staff to enable them more effectively to sort out and utilize the input of data, proposals, and suggestions that may be expected to flow in increasing volume from scientific, technical, professional, and industrial groups, from government agencies, and from the general public.

It may be useful to add a brief consideration by types of the various means available within the political process to help effect desired remedial changes. For this purpose the political process is defined to include the legal system. The order of listing is not intended to reflect relative importance. There are several means available:

1. Governmental measures to eradicate or diminish pollution caused by governmental operations, military or civilian. This can extend to mixed governmental-private operations and to private operations largely supported by governmental funds.

2. Taxes designed as incentives, stimuli, or pressure, together with taxes intended to provide revenue to finance remedial measures and research—such as effluent taxes; accelerated write-offs of investment in new technology intended to reduce pollution; excise taxes on the sale or purchase of materials or products that yield pollutants, with the revenues perhaps allocated to remedial measures or research.

3. Governmental financing of research into critical environmental problems and remedial technology.

4. The provision of financial assistance by governments out of general revenues to facilitate adjustments to remedial changes.

5. Common-law remedies, incrementally adjusted to felt contemporary needs, administered through the courts, for instance, actions for a nuisance, products liability, liability for harm caused by abnormally dangerous activities.

6. Regulation—in the form of either a statute administered through the courts or, more typically, a statute, and administrative agency, and supplementary action through the courts.

4.3.3

Illustrative Applications

The general overview must be tested by attempts to apply it to particular residuals emitted from particular types of sources (or protosources or secondary sources). The model can be given full, concrete, and operational meaning only by such an application in thoroughgoing detail. A few participants undertook to sketch such applications in summary and tentative form on the basis of their prior experience and currently available data. They did so for illustrative purposes, as a guide to what must be done. Some of these studies will appear in the technical volumes to be published.* We believe it will be helpful to present selected parts of certain of these studies at this point. We stress that these are merely selected parts of studies that are themselves tentative and summary and that they are presented for illustrative purposes only to illuminate the manner, problems, and possibilities of particular applications.

---

* The studies will relate to such pollutants as toxic metals, DDT, oil, and phosphates, and to such sources and protosources as the automotive industry and electric power companies.

Phosphates

Phosphate is frequently the limiting nutrient for the growth of algae in estuaries as well as in lakes and rivers. An increase in the aquatic phosphate level over that naturally present will often accelerate the rate of eutrophication. It appears that total phosphate discharges in the United States into waterways exceed those of 50 years ago by a factor of 3 or 4. In consequence the eutrophication of some rivers and lakes has reached the point where it is evident even to laymen. It is estimated that the emission of phosphates from cities and from farms aggregates over 1 million tons annually. In the Lake Erie basin, the estimated breakdown is rural runoff, 17 percent; urban runoff, 7 percent; industrial discharges, 4 percent; and discharges into sewers, 72 percent (Federal Water Pollution Control Administration, 1968). Of the discharge into sewers, a large portion derives from phosphate detergents. This pattern is probably not significantly different from that of the nation as a whole, except for the runoff from animal feed lots, which are a negligible factor in the Lake Erie basin but important elsewhere. Reliable figures for such runoffs are not available (Federal Water Pollution Control Administration, 1970).

Control techniques now available or likely to become available in the immediate future can be used to reduce runoff from farms as well as industrial discharges, to curtail or eliminate the phosphate in detergents, and to remove most of the phosphate from sewage.

1. For farms, better soil conservation practice can be instituted to limit surface erosion. The practice of spreading fertilizer or manure on frozen fields can be discontinued. The emission of wastes from animal feedlots and chicken farms can be reduced through procedures for effluent treatment or possibly recycling. Together, such techniques could be helpful in many local situations. On a nationwide basis, however, they could be expected to remove only a small proportion of the existing phosphate discharges.

2. To curtail industrial phosphate discharges, processes can be instituted similar to those available for municipal wastes. On a nationwide basis such treatment would again affect only a small proportion of the aggregate national emissions of phosphates.

3. The manufacture of detergents containing phosphates appears

to be a prime target for remedial action. The obvious procedure would be to seek to substitute for phosphate-type detergents other cleaning agents that are biodegradable and nontoxic. At present nitrilo-triacetic acid (NTA) appears to offer the highest promise. Opinions concerning its effectiveness differ. There appears to be general agreement that it would be reasonably effective, but some believe it would not achieve comparable cleanliness in the washing of clothes or dishes. It appears to be not quite as active as polyphosphate, and some details concerning its biological effects remain to be ascertained.

4. To the extent that phosphates are not removed from detergents, it might nevertheless be possible to keep them from reaching estuaries, rivers, or lakes by removing them from sewage. Such treatments would also remove phosphate in sewage derived from human wastes. If viewed in terms of municipal responsibility for the treatment of sewage, a sewage treatment plant might be regarded as a source; from another point of view, it may be considered a route; we here refer to it as a secondary source. Procedures are available to remove phosphates from the sewage by precipitating it from primary- or secondary-treatment effluents by using either lime (effective to remove 80 to 90 percent) or iron or aluminum salt (90 to 95 percent effective for removal purposes). The cost of such treatment appears to run at about 5 cents per thousand gallons (American Chemical Society, 1969).

It remains to consider how the available techniques can be brought into use.

1. In the case of farmers other than the larger animal feedlot operators, regulatory measures would appear to be unpromising. The administrative burden would be far out of proportion to the net gain that might be anticipated in nationwide terms. In the case of large animal feedlot operators, it may be practicable to control the discharge of phosphate-rich manures into streams by regulation. Kansas is currently testing such an approach through sanitary regulations issued by the state Health Department. Otherwise, education and dissemination of knowledge seem to be the best alternatives.

2. In the case of industrial discharges of phosphates, the prospects for regulatory measures appear to be somewhat less forbidding than in the case of farms, but not much. Here again, one

could question the practical return to be expected from the issuance of water quality and emission standards for phosphates applicable to industries and their enforcement, having in mind the relatively small proportion of phosphate pollutants attributable to this source.

3. As previously indicated, the main targets for action would appear to be the manufacture of detergents and municipal sewage treatment. An imaginative combination of excise taxes with grants-in-aid for municipalities financed from such taxes might be a suitable way of dealing with both. At the present rates of use, it is estimated that an excise tax of some 5 cents per pound on the use of phosphates in detergents should suffice to stimulate procedures for phosphate removal as well as a search for effective substitutes. The proceeds might be turned to grants-in-aid to municipalities to enable them to institute the available techniques for the removal of phosphate from their sewage. The incentive to municipal action could be increased if the tax were made slightly higher, enough to give the municipalities a small profit.

A comprehensive cost-benefit analysis in social terms of the foregoing measures would indicate (1) as benefits, reduced eutrophication, with a corresponding improvement in the aesthetic and recreational qualities of lakes, rivers, and estuaries; (2) as costs, a higher price for detergents, and perhaps somewhat reduced cleanliness in dishes and clothes washed in automatic dishwashers and washing machines; (3) as uncertainties, the effect on fish production. There is some evidence that reduced phosphate discharges into aquatic reservoirs might result in a somewhat smaller quantity of fish production, perhaps counteracted by the better quality of the fish that would be available.

Oil in the Oceans

(See also reports of Work Groups 2 and 5.) Oil enters the oceans from many sources in many ways. For our immediate illustrative purposes, only the discharge of oil from tankers in their normal operations and oil pollution derive from spent motor oils will be considered.

TANKERS. Tankers in their normal operations (not considering catastrophic losses) are estimated to account for some 530,000 metric tons per year of oil discharged into the oceans. Of this, approximately 500,000 metric tons are derived from tankers in the world's fleet that do not employ a simple and relatively low-cost control technique known in the trade as "Load on Top" (LOT). Tankers using LOT are responsible for only 30,000 metric tons. Yet the tankers using LOT constitute 80 percent of the world's fleet, while those that do not, represent only 20 percent. If the 20 percent not using LOT should adopt it, their discharges of oil through normal operations would be reduced from 500,000 metric tons per annum to 7,500 metric tons per annum, and the aggregate of oil discharged into the world's oceans from this source would be correspondingly reduced from 530,000 metric tons per annum to 37,500 metric tons per annum (see report of Work Group 5 for source and amplification of figures). In technological terms the appropriate remedial measure appears to be clearly indicated. The difficulty lies not in the available technology but in the means of bringing about its use.

It is necessary to point out that in considering pollution by oil in the oceans, we have passed from environmental problems for which remedial action can realistically be considered in national terms to environmental problems for which remedial action must be considered in international terms. In doing so, we anticipate a later section.

An international organization exists which has jurisdiction to examine such problems. As recently as November 29, 1969, two new international conventions relating to oil on the high seas were agreed upon and open for signature under the auspices of a conference of the Inter-Governmental Maritime Consultative Organization. One is entitled *An International Convention Relating to Intervention on the High Seas in Cases of Oil Pollution Casualties,* and the other is the *International Convention on Civil Liability for Oil Pollution Damage.*

The governments of forty-nine states were represented at the conference. Nineteen of these, including the United States, the United Kingdom, France, and Italy signed the Conventions' pending ratification in accordance with their respective constitutional processes. Twenty-nine did not. The nonsignatories included the Soviet Union, Japan, Norway, Sweden, Denmark, The Netherlands, Greece, and Liberia. Panama and Honduras, the other members of the so-called PanHonLib group,* appear not to have taken part in the Conference at all.

The extent of nonparticipation in the Brussels Conventions of November 1969 suggests the difficulties that might confront an attempt by international agreement to compel tankers to install LOT. The attitudes of the various states heavily involved in shipping are shaped by their respective preoccupations and priorities, which differ markedly. It does not necessarily follow that unilateral action by single states to compel the adoption of LOT, or joint action by several states, would be doomed to ineffectiveness. If the United States, for example, which plays so large a role in the oil market, were to ban tankers not using LOT from its ports, the resulting pressure on tankers to adopt LOT might be considerable, and such pressure could be intensified by parallel action on the part of the United Kingdom and other states. The pros and cons of such a possible ban would have to be considered by the United States not only in terms of its concern for the environment but also in the perspective of its trade policy and foreign policy. We believe that the possibilities of action along such lines should not be excluded by the United States from its survey of available choices.

SPENT MOTOR VEHICLE OILS. Spent motor vehicle oils in the aggregate may constitute a significant component of oil pollution in the oceans. Since complete data are not available to indicate the amounts that actually find their way into the world's waters, figures for spent motor oils must be used with caution. A reasonable estimate appears to be that on the order of 500,000 metric tons of spent motor vehicle oils go into the oceans annually. This is the cumulative effect of such activities as changing the oil in an engine (see report of Work Group 5 for source and more detailed discussion of figures).

The sources of spent motor vehicle oils—owners of passenger automobiles and trucks, filling stations, garages—are far too nu-

---

* The PanHonLib group is a term applied by the trade to ships registered in, and flying the flags of, Panama, Honduras, and Liberia. These are substantial fleets. The actual owners are in large part nationals of other countries, including the United States. The tanker, *Torrey Canyon*, whose wreck on rocks in the high seas off the coast of England in March 1967 spilled 117,000 tons of oil into the water, was registered in Liberia and flew the Liberian flag; the ship was owned and chartered by U.S. nationals and manned by an Italian captain and crew.

merous and widely distributed to make regulation a practicable option. The difficulties of enforcing compliance are compounded by the absence of any incentive for the individual source to take special measures in the course of disposal. It appears necessary to devise new institutions to afford any realistic prospect of effective remedial action. Such a new institution might take the form of a system of collection facilities suitably distributed, to which spent motor oils could be brought for appropriate treatment and disposal. The collection facilities would have to be supplemented by a system of incentives, the simplest form of which might entail payment by the facility for spent motor oil in an amount sufficient to induce the sources to bring it in. The payments might be financed by an excise tax on the sale of motor vehicle oil, to be levied in the first instance on the refiner or distributor, who would pass it on.

A problem would remain concerning the ultimate disposition of the oil at the collection facilities. Reprocessing appears to be technically feasible and to some extent at least might be economically practicable.

Remedial arrangements along the foregoing lines would appear to be manageable only if sufficient motivation to adopt them could be established to call forth a response from the political process. The necessary legislation in itself would not be hard to design; the required administrative organization in itself would not be hard to build; the budgetary cost in itself would not be great; the change in the habits of motorists in itself would presumably not cause too irritating a wrench. In the aggregate, however, the effort they would require would not be negligible, and it would have to overcome inertia. Current public and political opinion in America appears to be seriously concerned about oil in the world's oceans, and, if the concern is maintained and possibly enhanced, it will be an indispensable help to those seeking remedial measures. To induce affirmative action, however, it will also be necessary to come forward with more reliable calculations than are now available and to persuade political and public opinion that spent motor vehicle oils require attention along with offshore drilling, discharges from tankers and other shipping, and catastrophes such as that of the *Torrey Canyon*.

Carbon Dioxide

We pause to review the broad terms of our analysis. This chapter deals with change and its implications: what should be done and what the doing may involve. The question of change arises in regard to a residual only if (1) the fact that it is a key pollutant with harmful global effects has been established with a sufficient approximation of certainty or degree of probability to warrant remedial action, or (2) informed scientific and professional opinion, or public and political opinion, or both, view it with sufficient apprehension or concern to warrant appropriate measures. For the purposes of the report of this Work Group, we have assumed that the residuals included in our roster of key pollutants either have met the first criterion or have met the second with a realistic prospect that a few more years of research and experience would also satisfy the first.

CO_2 introduces an element not previously mentioned, relating to the scale and intensity of the possible effects. In the usual case, if there appears to be only a remote and highly speculative possibility that a residual might have harmful global effects, little time and effort will be put into a program of inquiry affecting it. However, if the speculative effects are of such a nature that they would be devastating if they should occur and if it would require long years of arduous preparation to afford a realistic possibility of achieving preventive or corrective measures, prudence might indicate that a serious program of inquiry should be instituted and sustained. CO_2 presents such a special case.

In 1861, Tyndall suggested that variations in the amount of carbon dioxide in the atmosphere would result in climatic changes through resulting variations of the surface temperature of the earth. In 1896, Arrhenius computed that a threefold increase of atmospheric carbon dioxide would cause a surface temperature rise on the earth of 9°C. Some contemporary investigators have suggested that an increase in the concentration of CO_2 in the atmosphere would warm the lower and middle troposphere and cool the stratosphere (see report of Work Group 1).

The present concentration of CO_2 in the atmosphere stands at a little over 0.03 percent by volume or 320 parts per million. It appears to be increasing at the rate of about 0.2 percent per year or 0.65 part per million each year (see report of Work Group

1). It is generally agreed that the increase is associated with man's mining and burning of fossil fuels—coal, oil, and gas—deposited millions of years ago. Various forecasts have been made as to what the aggregate increase in atmospheric carbon dioxide may amount to by the year 2000, but such forecasts are subject to major uncertainties relating to the rate of fossil fuel consumption, the nature of CO_2 reservoirs in the ocean and in the biosphere, and the effects of possible deforestation in connection with the development of heavily wooded areas, especially in the less-developed countries. The actual effects of an increased CO_2 concentration in the atmosphere remain highly problematical and the subject of considerable division in the scientific community. These questions are discussed more fully in the report of Work Group 1.

If it should turn out that the increase of the concentration of CO_2 in the atmosphere over the next few decades would bring clear indications of an incipient global climatic change, the consequences for the human condition and human endeavor could be enormous. They could threaten man's agriculture and food supply, his warmth in winter and his cooling in summer, and could throw his entire transportation system out of gear. A radical curtailment of man's consumption of fossil fuels would be required. The scale and intensity of the actual effects and the scope and nature of the necessary response would be greatly affected by the time estimated to be available for corrective action.

It is, furthermore, hard to conceive of an effect more authentically global than an effect on the world's climate, and corrective action to be effective would have to be correspondingly universal. It is not hard to imagine the bitterness and recriminations that might be injected into international relations by mutual suspicions concerning the scale and pace of the reduction in the consumption of fossil fuels in different countries. The requirements of the occasion would test to the limit mankind's political and administrative capacity to establish and manage international controls.

In putting forward the foregoing analysis, it is not our intention to enter the sphere of doomsday prediction. We believe that extensive research and systematic monitoring should be undertaken in relation to CO_2. The monitoring institutions and activities should be established on an international basis. Time, effort, and money will be required. For reasons explicit and implicit in the foregoing analysis, we believe that the necessary expenditures should be made.

4.4
International Action and Reaction

Our focus on global effects has given it an international orientation. The geographical distribution of sources and the practical requirements of research and remedial action have imported a national orientation. The principle of presumptive responsibility of sources, protosources, and secondary sources points to industrial enterprises, electric utilities, farms, and homes within nations, and to municipal and national governments. From the nature of the phenomena, corrective action to eliminate, diminish, or counteract the effects of key pollutants must be taken largely on a national basis.

Typically, however, remedial measures by single nations will have to be supported by parallel undertakings in other nations. Frequently, the nature of the problem will require collaborative international action in support of the separate actions within the several states. Situations may also arise in which remedial measures taken by one nation in the conviction that they are necessary and therapeutic will be resented by another nation and provoke a political reaction. Resentment will spring from consequences for the second state that are adverse in fact or strongly believed to be adverse. The political reaction can be as intense in the one case as in the other, and practical remedial measures may have to be supplemented by complex international adjustments.

4.4.1
Joint International Action

We have already touched upon two key pollutants—oil and carbon dioxide—for which optimal or even satisfactory results depend upon parallel and joint international action. National differences in values, outlook, geographical location, economic condition, technological capacity, and appraisals of data and evidence impose limitations upon such international action. A realistic appreciation of the obstacles is not inconsistent with the will to surmount them or with a conviction that they can be surmounted in a reasonable measure in time by imagination and tenacity. The difficulties may perhaps be viewed as a projection onto an international backdrop—with a corresponding multiplication of scale, intricacy, and intensity—of the problems previously examined in this section attending attempts to infuse corrective attitudes and corrective measures into the market and the political process within a single nation.

An awareness of the problems and possibilities, or at least preliminary indications of an awareness, have begun to appear within the United Nations, the Organization for Economic Cooperation and Development (OECD), and other existing international organizations and agencies. Since we believe a realistic appreciation of the difficulties to be a necessary condition to any prospect for success in coping with them, we should perhaps note our present view that to date the inventory of existing organizations is far longer in names than in demonstrated resources, will, or appreciation of the functions to be performed or even of the problems and possibilities.

Some international measures have been instituted involving steps beyond the acquisition and analysis of data. The Inter-Governmental Maritime Consultative Organization (IMCO), which was established in 1957, has previously been mentioned in this chapter. Following the *Torrey Canyon* disaster, conferences were organized in London in May 1967, at the headquarters of IMCO, to consider whether and how progress might be made in regard to pollution of the seas by oil. The discussion covered such matters as the designation of sea lanes for tanker traffic, navigational aids, shore guidance systems, construction and design, and the demarcation of prohibited areas. IMCO was requested to make further studies, which were followed by the Brussels conference and the preparation and opening for signature of the two conventions already mentioned. *The International Convention Relating to Intervention on the High Seas in Cases of Oil Pollution Casualties* permits parties to the Convention to take prescribed measures on the high seas to protect their coasts from oil pollution. In general, the corrective measures must be reasonable and proportionate to the threat of harm to the coasts of the intervening state. Casualties involving warships are excluded. *The International Convention on Civil Liability for Oil Pollu-*

tion Damage imposes strict liability on ships of parties to the Convention for damage to the territorial waters and coasts of other parties caused by discharges of oil into the sea in the course of shipping operations. The Convention also contemplates that IMCO will draft and submit for approval a compensation plan to cover cases inadequately treated by the Convention itself. Both treaties are too new to make it possible to judge how widely they will be accepted or how effective compliance will be on the part of those nations that do accept it.

In regard to radioactive wastes, the activities of the International Atomic Energy Agency (IAEA) have consisted essentially of the collection, exchange, and general dissemination of information and the publication of suggested guidelines. It has published a guide book on radioactive waste management as well as on aspects of safety in nuclear operations not directly related to environmental pollution. In the current year it is arranging for panels of experts to examine problems relating to the control of certain radioactive components of airborne wastes, and it plans to hold symposia on the environmental implications of nuclear power stations. It has set up in collaboration with the World Health Organization and UNESCO a worldwide sampling network to give some measure of the precipitation of radioactive pollutants, and it maintains at Monaco an International Laboratory of Marine Radioactivity to conduct research that extends to problems of waste disposal and the standardization of techniques to determine the effects of radioactivity in the sea.

The only significant step going beyond the collection and analysis of data in relation to radioactive pollutants has been taken outside the sphere of the International Atomic Energy Agency and has been predominantly motivated by a concern for arms control. We refer, of course, to the nucelar test ban treaty. In the degree to which nations adhere to the treaty and comply with its terms, there will be a corresponding reduction in what has been to date the chief risk of radioactive contamination of the environment. The treaty represents a notable achievement, but the failure of states such as France and Mainland China to accede to it may be taken as one more example of the difficulties that attend such an international undertaking.

Given the fact that the shift in values and the distribution of priorities between the first-order effects and the side effects of technology is of recent origin and has been under way for only a brief period even in the United States, given also the uncertainty of much of existing knowledge concerning key pollutants and their effects, and bearing in mind the complexity of international arrangements, it is not at all surprising nor should it be discouraging that the record of international regulatory measures affecting pollution should be rudimentary. It is in the sphere of the collection, organization, and analysis of essential data that the prospects are best for meeting the burden of demonstration and persuasion needed to engender a response in the international political process. Accordingly, in a separate report on monitoring, some preconditions for an international monitoring network have been discussed.

4.4.2
National Action and Reaction: Pollution and
the Developing Countries
Spreading from their places of origin in the seventeenth and eighteenth centuries in Europe and its outposts in North America, the scientific and industrial revolutions have penetrated nearly every corner of the planet. The penetration has been uneven, as have been the elements of indigenous scientific and technical innovation that preceded or followed it. In the current stage of this immense and continuing process, the disparities in the rate and depth of change remain such that one-third of the world's population extracts several times as much beneficial use per person from the environment as the other two-thirds. Precise measurement of this inequality is hindered by problems ranging from statistical difficulties to deeply felt differences about what constitutes a benefit to man and what must be entered on the other side of the ledger as a corresponding cost. Nevertheless, there is no doubt that a large gap between the advanced industrial societies and the less-developed countries (LDCs) exists, and there is much evidence to suggest that it is widening and that it may continue to widen through the end of the present century.

It is little solace to the seeker of reasonable order in the world that the gap between rich nations and poor may have less influence upon aspirations and behavior than the contrast in the mind of a citizen of a developing country between the way his father lived and the way he wills that his children shall. The growth of inexpensive and universal communications among peoples is steadily eroding the barriers that have for most of history assured that the great majority of mankind would conceive of the conditions that ought to prevail in their societies as some plausible function of the conditions that do prevail in such societies. To a large degree their sense of what ought to prevail has become a function of conditions observed elsewhere in the world. In consequence, the gap between desire (reflected in political demand) and current reality in developing countries probably corresponds in scale and rate of increase to the gap in standards between the advanced industrial societies and the LDCs.

The situation is replete with political, economic, and social implications of profound significance, most of which lie outside the scope of this analysis. In no respect, however, is the contrast in interests, priorities, and capacities between the rich states and the poor more stark than in dealings involving environmental pollution. All the ingredients are present for a classically unproductive dialogue of the deaf, and the loose and half-hearted sparring that has taken place to date fits all too well into that category. Perhaps this cannot be helped. Perhaps joint or parallel action between the rich and the poor on the environmental front cannot realistically be expected. Nevertheless, the values at stake are worth the best effort we can muster to cut through the web of cross purposes and achieve some measure of cooperation or at least of understanding. And the first step is to recognize that there is almost no resemblance between the predominant views of the environmental problem on the two sides of the income gulf.

The basic differences in view do not primarily reflect differences in the reach or intensity of environmental deterioration. It must be borne in mind how large a portion of the people in developing countries live in cities (the concentration of the populations of Mexico and Argentina in Mexico City and greater Buenos Aires are illustrative), and urbanization in the LDCs is proceeding more rapidly than in the advanced industrial societies. India's industrial plant matches Italy's in size and probably generates substantially more uncontrolled pollutants. No one who has seen the air above Sao Paulo, or the waters of the Hooghly as it flows through Calcutta, or the coastal waters of West Africa can retain an illusion that the poor countries do not have pollution. The pollution in fact is comparable in intensity to that in the advanced industrial societies. Some of the LDCs are also beset by forms of pollution that are not experienced in the rich countries and are caused by the sheer numbers of peoples and animals.

Since the populations of the poorer countries are much larger than those of the rich and exist on a much narrower margin of subsistence, an external observer, taking what he considers an objective view, would conclude that such populations have at least as great a stake in general effects of pollution as their better-off brethren. If the albedo (reflectivity) of the earth is changing because of man-generated particles, if the CO_2 content of the atmosphere is rising, if the mean temperature is changing, if the oceans are being contaminated with oil or radioactive wastes, or indeed if any significant and potentially harmful change in the environment is occurring, the less-developed countries have every reason—in the kind of objective sense that would mark the judgment of an external observer—to be just as alarmed as the developing countries and just as ready to take strong and costly action. Yet they are not.

The difference lies not in the physical facts of pollution but in the attitude with which the facts are received, an attitude toward first-order effects and side effects much closer to the mood and perception of early nineteenth-century America than to the contemporary American outlook. The LDCs tend to regard the rising concern about environmental deterioration among the wealthy nations as at best an irrelevance and at worst a plot to cheat them of their rightful share of the world's resources and of the continuing promise of the scientific and industrial revolutions. The discussion of environmental issues between the rich countries and the poor cannot be fruitful or even meaningful until the former recognize (1) that this is in fact the standard view of the matter in the developing countries, and (2) that the view is not necessarily unreasonable when seen either from their standpoint or from the standpoint that prevailed in the richer countries earlier in their own histories.

It may facilitate such a recognition to reexamine the history and nature of our own shift in values and to sort out and examine separately some of its components, along with some of the forces that have shaped it. Such a shift in values is an immensely complicated and extremely subtle phenomenon, and it would be foolhardy to assume that one can fully comprehend it, not to speak of compressing an analysis of it into so brief a compass. Nevertheless, it may not be amiss to attempt to suggest a few of the factors involved.

The people of the United States—as of other advanced industrial societies—have recently acquired a new appreciation of the scope of the actual and potential effect of human activities on the environment. The appreciation is based partly on new facts and partly on new habits of mind engendered by the new facts, the continuing record of massive technological achievement, and other elements of the national experience too subtle to be precisely identified. The habit of mind involves a belief by the ordinary man that the collective technological capacity of his society is such that it can alter the balances of the environment either to the benefit of mankind (control the weather) or to mankind's ultimate, fatal detriment (the doomsday outlook). It would be uncommonly difficult to explain such a belief to the ordinary inhabitant of one of the LDCs. The mind of the latter simply does not accommodate a concept that the actions of his people, conscious or unconscious, will radically affect the fate of the planet itself.

Another factor to be taken into account is a widespread apprehension in the United States that the ultimate supplies of natural resources available and necessary for our immense and specialized technology may be exhausted. Whole generations have grown up reading that at present rates of consumption the earth's stock of oil (or bauxite or copper or iron) would be played out by some date calculated through imaginative mathematics. The dates tend to move onward with the flux of events, but a general sense persists that the world is consuming its high-quality resources at an accelerating rate and that prudence indicates the wisdom of conservation. The apprehension concerning the potential exhaustion of resources is psychologically related to widespread feelings that such recent additions to life as air conditioning or television have become virtually necessities. In the LDCs, by contrast, the notion that human activities can exhaust any feature of the planet has not taken root. The prevalent attitude with regard to natural resources is that they should be exploited ever more rapidly with ever greater efficiency within the originating country. In addition, a large proportion of the population of the LDCs is not so far removed from the standard of living of preindustrial life as to be appalled by the prospect of a partial return to it.

The flux in values in the advanced industrial societies also contains a more complex component involving the philosophical concept of the good. In contemporary discussion, suggestions are not uncommon that the national income in the aggregate is sufficient for reasonable standards, that the importance of sustained continuing economic growth has been exaggerated, and that the national consciousness should be focused upon the quality of life. If much of this reflects current fads, much also reflects a genuine travail of spirit. But it is essentially incomprehensible in the LDCs. In the developing countries, economic growth and material progress are not objectives among other objectives to be emphasized or deemphasized at will or at whim. They constitute the predominant objective by which others are largely measured. The developing lands simply do not see the quality of life as a trade-off against economic growth.

The foregoing analysis may perhaps convey some sense of the potentialities for misunderstanding and confusion between the advanced industrial societies and the LDCs in an international conference called to deal with environmental problems and pollution. There is little reason to believe that the developing countries can be diverted from their preoccupation with the first-order effects of technology to a concern about the side effects upon the environment. Currently, and in the foreseeable future, the advanced industrial societies will have to carry the load of remedial action against pollution. When they need or desire the cooperation of the LDCs, they will be able to obtain it, if they can do so at all, only by paying for it. The payment may have to cover considerably more than the costs of research or of the management of antipollution measures. It may involve the incorporation of a new component into programs of aid to the LDCs, in the form of compensation for extra costs or competitive disadvantages injected into the development programs of the LDCs by antipollution measures.

Earlier in the present chapter, we referred to situations in which antipollution measures adopted in one country with beneficial consequences may have effects in another country that are adverse in fact or deemed adverse in the second country. Measures relating to DDT afford a striking example. The concern of the United States and other advanced industrial societies to curtail the use of DDT and counteract the consequences of its prior use strikes no responsive cord in the LDCs. They are apprehensive lest they be deprived of DDT, and they are making their apprehensions clear. On the present evidence, nothing will reconcile them to such a deprivation except a substitute at least as effective, safe, easy to handle, and inexpensive as they have found DDT for the purposes to which they attach primary importance, notably antimalarial and other health campaigns and improved yields in their agriculture. A study exploring the relations between the advanced industrial societies and the LDCs in terms of DDT will be incorporated in one of the technical volumes.

4.5
The Function of Scientific, Technical, and Professional Education

In conclusion, we believe it appropriate to stress the significance of scientific, technical, and professional education and training. The development of science over the past three centuries, the application of science in technology, and the application of technology to economic and social life through new forms of economic and social organization have been a function of new knowledge and skills embodied in scientific, technical, and professional personnel. The scientific and industrial revolutions consist of the cumulative effect of the products of the minds and hands of such personnel. Their fundamental significance can be observed in the developing countries. Not only is virtually all the important technology imported but also, in many cases, the manpower required to explain, install, and use it. The ultimate factor in the modernization of an LDC will be the bringing into being within its own population of a cadre of such personnel sufficient in numbers and quality to the size and circumstances of the LDC.

Such people made possible the first-order effects of technology. Such people over time can make the critical contribution to control over the side effects of technology.

A concern for the side effects of technology, an understanding of the interrelationships among the side effects and the first-order effects, and a constant disposition to explore the full range of technological possibilities to realize the optimum potentialities of technology on a balanced basis should be incorporated into the regular education and training of scientists and engineers, as well as of lawyers, economists, and other professional and technical groups. In a sense this is obvious, but we believe it to be an area in which the obvious must be emphasized and freshly reabsorbed.

References

American Chemical Society (ACS), 1969. *Cleaning Our Environment: The Chemical Basis for Action* (Washington, D.C.: ACS).

Federal Water Pollution Control Administration (FWPCA), 1968. *Lake Erie Report* (Washington, D.C.: U.S. Government Printing Office).

Federal Water Pollution Control Administration (FWPCA), 1970. Animal wastes profile, in *The Economies of Clear Water*, Vol. II (Washington, D.C.: U.S. Government Printing Office).

Panel on Technology Assessment of the National Academy of Sciences, 1969. *Technology: Processes of Assessment and Choice,* a Report prepared for the Committee on Science and Astronautics, U.S. House of Representatives (Washington, D.C.: U.S. Government Printing Office).

5.
Work Group on Industrial Products and Pollutants

Chairman
Raymond F. Baddour
MASSACHUSETTS INSTITUTE OF
TECHNOLOGY

Robert U. Ayres
INTERNATIONAL RESEARCH AND
TECHNOLOGY CORPORATION

Edward Corino
ESSO RESEARCH & ENGINEERING
COMPANY

Dale W. Jenkins
THE SMITHSONIAN INSTITUTION

Milton Katz
HARVARD LAW SCHOOL

Thomas Marqueen
BOISE CASCADE CORPORATION

George B. Morgan
NATIONAL AIR POLLUTION CON-
TROL ADMINISTRATION

James T. Peterson
NATIONAL AIR POLLUTION CON-
TROL ADMINISTRATION

Henry Reichle
NATIONAL AERONAUTICS AND
SPACE ADMINISTRATION

Frederick E. Smith
HARVARD UNIVERSITY

Rapporteur
Jonathan Marks
HARVARD LAW SCHOOL

5.1

Introduction

This Work Group report presents data on several products of industrial societies that are known or potential pollutants (pesticides, toxic heavy metals, oil, and phosphorus). It also presents estimates of the amounts of selected pollutants that enter the environment as a direct result of industrial processes (gaseous and particulate emissions).

In most cases the only well-documented data available to us were for the United States. For other parts of the world we recommend that existing data on relevant products and processes be compiled and that new data be gathered for a more complete picture. It is especially important for environmental forecasting that projections of future industrial growth and product demands be developed.

5.2

Production and Use of Pesticides

U.S. production of pesticides, which includes insecticides, herbicides, and fungicides, has steadily increased up to last year when it fell 7.4 percent (U.S. Tariff Commission, 1970). The data on U.S. production of synthetic organic pesticides shown in Table 5.1 indicate an average annual growth rate of 10.5 percent for the 1964–1968 period.

Table 5.1
U.S. Production of Synthetic Organic Pesticides[a]
(Thousands of metric tons)

Year	Production[b]
1960	294
1962	332
1964	356
1966	457
1968	545
1969	505[c]

Ten-year production total, 1959–1968 = 3.79 million metric tons.
[a] Includes insecticides, fungicides, and herbicides.

[b] Source: USDA, *Pesticide Review*, 1969.
[c] Source: U.S. Tariff Commission, 1970.

Accumulating evidence indicates that persistent pesticides are contaminating the environment and adversely affecting living organisms (see report of Work Group 2). These persistent compounds are typified by the chlorinated organic insecticides such as DDT and the aldrin-toxaphene group. U.S. production and consumption of these materials are shown in Tables 5.2 and 5.3. These data indicate that U.S. use of DDT is decreasing while consumption of aldrin-toxaphene materials has been relatively constant.

Reliable information on worldwide use of pesticides was not readily available, but we believe world production to be no more than twice U.S. production.

5.3
Production, Use, and Emission of Selected Toxic Metals
Certain heavy metals are highly toxic to plants and animals, including man. They are highly persistent and retain their toxicity

Table 5.2
U.S. Production and Consumption of DDT
(Thousands of metric tons)

Years	Production	Consumption
1950	35	26
1952	45	32
1954	44	20
1956	63	30
1958	66	30
1960	75	37
1962	76	30
1964	56	23
1966	64	21
1968	63	15

Total production, 1944–1968 = 1.225 million metric tons.

Ten-year production, 1959–1968 = 676 thousand metric tons.
Source: USDA, *Pesticide Review*, 1969.

Table 5.3
U.S. Production and Consumption of Aldrin-Toxaphene Group[a]
(Thousands of metric tons)

Year	Production[b]	Consumption[e]
1956	39	28
1958	45	36
1960	41	34
1962	48	37
1964	48	38
1966	59	39
1968	55	18

Ten-year production total, 1959–1968 = 493,000 metric tons.
[a] Includes aldrin, chlordane, dieldrin, endrin heptachlor, strobane, toxaphene.

[b] Source: *Chemical Economics Handbook*, 1969.
[e] Source: USDA, *Pesticide Review*, 1969.

for very long periods of time. Some have been used extensively as pesticides or biocides and have been dispersed into the environment as pesticides, as uncontrolled industrial wastes and emissions, and by other means.

The main route into natural water systems for such materials is industrial waste discharge, often through municipal sewage systems. Identification and tracking of these metals in sewage and sewage sludge is difficult because of low metal concentrations and high concentrations of organic matter. Conventional sewage treatment removes only a portion of the metals from sewage.

A general strategy to limit the efflux of heavy metals in sewage is to eliminate their sources instead of removing metals by sewage treatment. Although there are scattered data on effluent metal concentration, overall mass balances of metals have not been made for sewage collection, treatment, and disposal systems.

There are about two dozen metals that are highly toxic to plants or animals. The most toxic, persistent, and abundant in the environment have been selected for special review. These include mercury (Hg), lead (Pb), arsenic (As), cadmium (Cd), chromium (Cr), and nickel (Ni). Most heavy metals are biologically accumulated in the bodies of organisms, remain for long periods of time, and function as cumulative poisons.

In order to attempt to assess the total environmental contamination sources, the total world production and U.S. consumptions are given for the period 1960 to 1968 in Table 5.4. The production and use data for arsenic were available only for insecticide use and are not presented.

Because of the attention given to mercury in the report of Work Group 2, more detailed information on mercury production and uses will not be given here.

The world production of mercury in 1968 was 8,810 metric tons. The U.S. production in 1968 was 1,000 metric tons from domestic mines, mostly from the states of California, Nevada, Idaho, and Oregon, 11 percent of the world production (*Minerals Yearbook*, 1968). The United States imported 827 metric tons in

Table 5.4
World Production[a] and U.S. Consumption[b] of Toxic Heavy Metals
(Thousands of metric tons)

	Hg		Cd		Pb		Cr$_2$O$_3$		Ni	
Year	World	U.S.	World	U.S.	World	U.S.	World	U.S.	World	U.S.
1960	——	1.77	——	4.53	——	930	——	1,110	——	98.2
1961	——	1.92	——	4.65	——	932	——	1,090	——	108
1962	——	2.26	——	5.56	——	1,010	——	1,030	——	108
1963	8.28	2.70	11.8	5.19	2,520	1,060	3,920	1,080	340	114
1964	8.81	2.81	12.7	4.31	2,520	1,090	4,150	1,320	372	134
1965	9.24	2.54	11.9	4.75	2,700	1,130	4,810	1,440	425	156
1966	9.51	2.46	13.0	6.60	2,860	1,200	4,390	1,330	414	171
1967	8.36	2.40	12.9	5.28	2,880	1,150	4,300	1,230	441	158
1968	8.81	2.60	14.1	6.05	3,000	1,200	4,730	1,200	480	144[c]

[a] Sources: 1963 data are from the *Minerals Yearbook*, 1967; 1964–1968 data are from the *Minerals Yearbook*, 1968.

[b] Source: *Chemical Economics Handbook*, 1969.

[c] Source: *Minerals Yearbook*, 1968.

1968. The total production and import amounted to 20.8 percent of the world merccry for that year (*Minerals Yearbook,* 1968; *Chemical Economics Handbook,* 1969).

From 1930 to 1970 the United States mined 31,800 metric tons, imported 39,600 (71,400 metric tons total), and exported 1,820, leaving 69,580 metric tons in the United States. Consumption during this 40-year period was 63,000 metric tons (*Chemical Economics Handbook,* 1969). A 40-year world production figure for mercury is unknown. The world use of mercury-containing pesticide reported by the FAO for the period 1948–1967 for agricultural use alone was 52,946 metric tons, not including the Soviet bloc or Korea (Food and Agriculture Organization, 1968).

The two major uses of mercury in the United States in 1968 were in electrical equipment and electrolytic preparation of chlorine and caustic soda. The major sources of mercury in the atmosphere are mining and refining processes, electrical manufacturing, cholrine and caustic processing plants, and scientific laboratories. Since mercury is expensive, much of it is collected and reused. The secondary recovery in the United States in 1968 was 1,190 metric tons, compared with total U.S. consumption of

Table 5.5
Gaseous Emission Factors

Industry	CO	SO_x	HC	NO_x
Wood pulp	0.03	0.0035	——	——
Nitric acid	——	——	——	0.029
Sulfuric acid	——	0.0213	——	0.029[a]
Smelter output	——	1.25	——	——
Oven coke	——	0.009	0.0172	——
Iron foundries	0.039	——	——	——
Petroleum products[b]	0.0756	0.053	0.064	——

Source: McGraw, forthcoming.

[a] NO_x is emitted in producing sulfuric acid only where the chamber process is used, about 3 percent of U.S. production. We assumed it was the same percentage of world production.

[b] This was originally in tons/10^3 barrels. To convert, we assumed all petroleum products were gasoline and that there were 8.50 barrels per metric ton (United Nations, *World Energy Supplies*).

2,600 metric tons; that is, 46 percent was reused (*Minerals Yearbook*, 1968; *Chemical Economics Handbook*, 1969).

Because of its high vapor pressure at room temperature, exposed mercury constantly emits vapors into the air. Any activity that heats mercury (or its compounds), such as mining and refining operations and mercury arc rectifiers, also emits mercury vapor. Pesticides (especially for fungus control), slimicides in the pulp paper industry, and paint are additional sources of environmental contamination.

5.4
Emissions from Industrial Operations

Table 5.5 is a summary of pollutant gaseous emission factors for some major industries that emit carbon monoxide, sulfur oxides,

Table 5.6
Summary of 1968 Worldwide Gaseous Emissions from Major Industrial Sources, Excluding Fuel Consumption[a]
(Millions of metric tons)

Industry	CO	SO$_x$	HC	NO$_x$
Chemically processed wood pulp	2.0	0.24	——	——
Nitric acid	——	——	——	0.52
Sulfuric acid	——	1.68	——	0.068
Smelter output[b]	——	16.2	——	——
Oven coke[c]	——	2.8	5.3	——
Iron foundries	15	——	——	——
Petroleum refineries[d]	9.17	6.4	8.1	——
Total	26.2	27.3	13.4	0.59

[a] Multiplied the world production figure for the industry (United Nations, *Statistical Yearbook*, 1969) and the emission factor (Table 5.5) to calculate the figures in the table.
[b] Smelter output includes smelter production of copper, lead, and zinc.
[c] Assuming the world produces oven coke only.
[d] To find the emissions for 1968, we multiplied the 1965 world production figure (American Petroleum Institute, 1967) by the petroleum products emission factor (Table 5.5) to get the 1965 emissions. To this we applied the growth rate of 7.7 percent calculated from data in American Petroleum Institute, 1967, by taking the ratio of 1969 production to 1959 production and determining the compound interest rate that would have yielded such growth. This rate was then used to extrapolate from 1965 to 1968.

Table 5.7
U.S. and Global Industrial Particulate Emissions, 1968
(10^6 metric tons)

Industry	1968 Global Production[a]	U.S. 1968 Percentage Global Production[b]	U.S. 1968 Uncontrolled Emissions[c]	U.S. 1968 Controlled Emissions[d]	U.S. Percentage Emissions Controlled[e]	Foreign 1968 Uncontrolled Emissions[f]	Global 1968 Uncontrolled Emissions[g]
Iron and steel gray iron foundries	915[h]	22	12.0	1.890	84	42.6	54.6
Grain handling, storage flour, feed milling	112[i]	10	2.3	1.020	56	20.7	23.0
Cement	513	13	7.9	0.790	90	53.0	60.9
Pulp and paper	90[j]	38	3.6	0.660	82	5.9	9.5
Miscellaneous[k]	—	—	8.0	2.465	69	—	—
Total			33.8	6.825	81		

Notes:
a All data in this column from United Nations, *Statistical Yearbook*, 1969.
b Derived from United Nations, *Statistical Yearbook*, 1969, by taking the ratio of U.S. production to world production.
c NAPCA Division of Air Quality and Emissions Data, unpublished data.
d NAPCA, 1970.
e $\dfrac{\text{Column 3} - \text{Column 4}}{\text{Column 3}}$

f $\text{Column 3} \dfrac{(100 - \text{Column 2})}{(\text{Column 2})}$

g Column 3 + Column 6
h Combination of figures for production of pig iron and crude steel.
i Includes only production of wheat flour.
j Combination of figures for chemical and mechanical production of wood pulp.
k See Table 5.8.

hydrocrabons, and nitrogen oxides. Given an industry's output, these factors can be used to calculate the tons of gaseous pollutants emitted. This was done for 1968 in Table 5.6.

Table 5.7 presents the estimates of the National Air Pollution Control Administration (NAPCA) of emissions of particles into the atmosphere by industrial processes in 1968. These numbers are important from an environmental viewpoint because atmospheric particles can attenuate both solar radiation and terrestrial infrared radiation and are thus able to affect climate on a global basis (see report of Work Group 1). The table is rather incomplete because the appropriate figures could not be found in the time available. We have estimates for U.S. controlled and uncontrolled particulate emissions and have derived estimates for world uncontrolled emissions (Table 5.8).

The heavy metals that are emitted into the atmosphere from industrial processes are important in that they are effective as freezing nuclei (see report of Work Group 1). Fe_2O_3 is significant

Table 5.8
U.S. Particulate Emissions for Miscellaneous Sources, 1968[a]
(10^6 metric tons)

Industry	Uncontrolled Emissions[b]	Controlled Emissions[c]
Sand, stone, rock, etc.	2.4	0.790
Asphalt batching	2.5	0.490
Lime	0.8	0.410
Phosphate	0.4	0.185
Coal cleaning	0.5	0.170
Other minerals	0.4	0.160
Oil refineries	0.2	0.090
Other chemical industries	0.3	0.085
Other primary and secondary metals	0.5	0.085
Total	8.0	2.465

[a] The industries for which there were global production figures are in Table 5.7.

[b] NAPCA Division of Air Quality and Emissions Data, unpublished data.

[c] NAPCA, 1970.

in this regard. Roughly 50 percent of the emissions from steel mills consist of Fe_2O_3, although there is considerable variation among different specific steel processes (Sullivan, 1969). The total emissions from this industry are included in the tables. The industrial emissions of lead are not significant on a large scale when compared to that resulting from automobile exhaust. Zinc in the form of ZnO also fits into this category; its main sources are zinc, lead, and copper smelters and various steel processes.

We were unable to find global estimates for the percentage of emissions controlled. There appear to be no general projections of industrial growth; therefore we have not estimated future emissions of pollutants. We recommend that these estimates be developed through the year 2000 if they do not already exist.

5.5
Industrial Wastes of Petroleum Origin

Oils of petroleum origin are sufficiently different from oils occurring naturally in living organisms that increasing discharge of

Table 5.9
World Crude Oil Production[a] and Transport by Tanker[b]
(Millions of metric tons per year)

Year	Production	Tanker Transported
1960	1,040	——
1962	1,210	——
1964	1,420	——
1965	1,500	——
1969	1,820	1,180
1975	2,700	1,820
1980	4,000	2,700

[a] 1960 to 1965 production taken from American Petroleum Institute, 1967. 1969 estimates from Esso Research and Engineering Company (Standard Oil of New Jersey). 1975 and 1980 estimates averaged between predictions in Bachman, 1969, and *Marine Resources and Legislative and Political Arrangements for Their Development*, 1969, assuming production equal to consumption.

[b] 1969 estimate from Esso Research and Engineering Company (Standard Oil of New Jersey). Estimates for 1975 and 1980 from *Oil and Gas Journal*, 1968.

Table 5.10
Estimates of Direct Losses into the World's Waters, 1969
(Metric tons per year)

	Loss	Percentage of Total Loss
Tankers (normal operations)[a]		
Controlled	30,000	**1.4**
Uncontrolled	500,000	**24.0**
Other ships (bilges, etc.)[b]	500,000	**24.0**
Offshore production (normal operations)[c]	100,000	**4.8**
Accidental spills[d]		
Ships	100,000	**4.8**
Nonships	100,000	**4.8**
Refineries[e]	300,000	**14.4**
In rivers carrying industrial automobile wastes[f]	450,000	**21.6**
Total[g]	2,080,000	**100.0**

[a] Sources: Inter-governmental Maritime Consultative Organization, 1965, and Esso Research and Engineering Company (Standard Oil of New Jersey). It is estimated that 80 percent of the world's tanker fleet use Load-on-Top to reduce discharges from cleaning and 20 percent do not. These estimates are based on this distribution and some measured values for oil content of discharge using both methods.

[b] Source: Moss, 1963. Firm data in this area are not available anywhere. Records are not kept of bilge pumping nor are estimates of oil content. This area warrants further development; surveys need to be run to produce further data.

[c] Sources: Surveys by Esso Research and Engineering Company (Standard Oil of New Jersey) and American Petroleum Institute, 1969.

[d] Sources: U.S. Coast Guard, 1969, and American Petroleum Institute, 1969. These data are not firm. Mechanisms are needed to accumulate data to improve the reliability of this estimate.

[e] Sources: American Petroleum Institute, 1968; Federal Water Pollution Control Administration, 1968; and Esso Research and Engineering

Company (Standard Oil of New Jersey). Petrochemical production wastes are included.

[f] Using U.S. Department of Health, Education, and Welfare, 1963, data, we have assumed that the average concentration of hydrocarbons in the U.S. rivers entering the sea is approximately 85 parts per billion. Multiplying this figure by the estimated river runoff of 1,750 billion tons/year yields a figure of approximately 150,000 tons of hydrocarbons, which are annually carried to the oceans from the United States (Revelle, 1963). If the river runoff for the world is three times that of the United States, and if the average concentration is the same as the estimated U.S. average, then about 450,000 tons of hydrocarbons are carried into the oceans. In addition hydrocarbons are introduced into the ocean through sewage outfalls. We estimate that this amount is less than that in the rivers perhaps by a factor of 3 or 4, but this is not included in the table.

[g] Oil from pleasure craft, whose use is increasing in the United States, is not included, nor are there any measurements of natural seeps, which would be valuable for comparative purposes.

Table 5.11
Selected Products and End Uses of Phosphate Rock

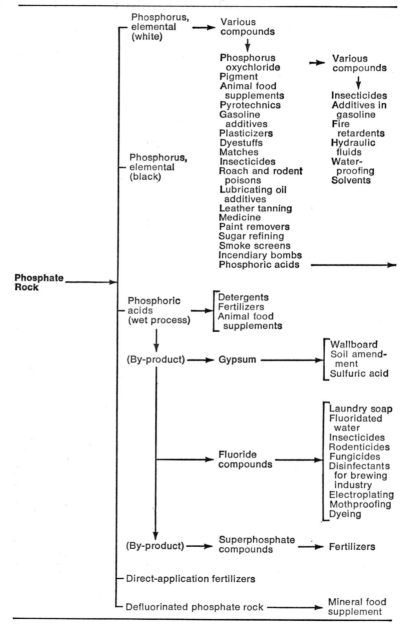

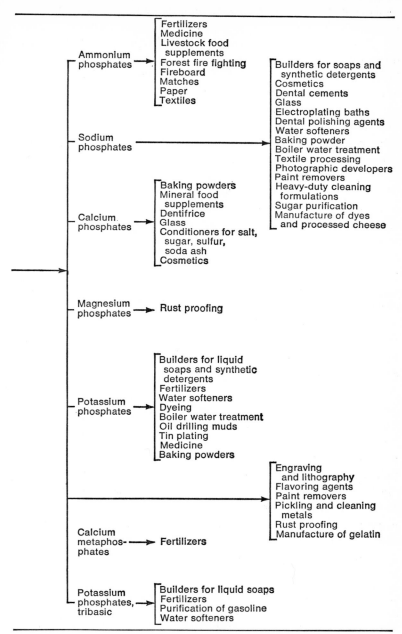

Source: Derived from Stanford Research Institute, 1967.

petroleum products into the environment may have serious ecological consequences, particularly in the marine environment (see
report of Work Group 2).

Table 5.9 outlines past and projected figures on world crude
oil production, illustrating how the production and use of oil has
increased and will increase. It includes projections of the amount
of oil tankers will carry in 1975 and 1980.

Table 5.10 details the various ways in which oil is introduced
into the world's waters. The relative importance of tanker operations as a source of pollution is evident from the figures. It is note-
worthy that a major contributor to oil pollution is the waste
lubricants from industrial and highway uses. It has been estimated
that between 500,000 to 1,000,000 metric tons of automotive lubricants are annually disposed of as waste (Glazier and Sumner,
1970). It is also possible that another 1,000,000 metric tons of
waste oil are generated by industry (American Petroleum Institute, 1970a). Very little is known about the degree to which the
disposal of these waste oils contributes to marine pollution. An
estimate can be made by looking at the concentration of hydrocarbons in river runoff (Table 5.10, fn. f). However, it is difficult
to make a reliable gross estimate of the amount introduced into
the marine environment through sewage outfalls because of the
wide variation in source characteristics, treatment processes, and
sludge disposal practices.

5.6
Phosphorus Cycling in the Environment

The report of Work Group 2 discusses the serious local, regional,
and global problems that can result from discharges of phosphorus into the world's waters. Yet reliable knowledge of the
amount of phosphorus mined by man that is ultimately released
to the environment does not exist. Even estimates are difficult because broadly accepted production figures do not even exist. Consider, for instance, the following disclaimer from the *Chemical
Economics Handbook,* 1969:

In attempting to arrive at a supply-demand balance [for phosphate rock] for 1966 based on "standard" methods, from 2–4
million tons (P_2O_5) of rock remain unaccounted for. This represents nearly 20% of total supply (on a P_2O_5 basis) and is more
than can be reasonably attributed to losses or variance in pro-

Table 5.12
U.S. Production of Phosphate Rock and Derivative Compounds
(Thousands of metric tons of P_2O_5 equivalent)

Ammonium phosphates	234
Calcium phosphate (monobasic)	261
Calcium phosphate (dibasic)	113.4
Direct phosphate rock application fertilizer	10.8
Elemental phosphorous (white)	**1,427**
Elemental phosphorous (red)	14.3
Ferrophosphorous	46.8
Fertilizers (from furnace phosphoric acid)	220
Fertilizers (from wet process phosphoric acid)	3,180
Normal superphosphate	1,060
Phosphate rock	7,650
Phosphoric acid (furnace)	1,110
Phosphoric acid (wet process)	3,530
Phosphorous oxychloride (from phosphorous pentoxide	5.7
Phosphorous oxychloride (from phosphorous trichloride)	19.4
Phosphorous pentasulfide	38.6
Phosphorous pentoxide	**1,119**
Phosphorous trichloride	25.8
Sodium acid pyrophosphate[a]	22.5
Sodium metaphosphate[a]	79.2
Sodium phosphate (monobasic)[a]	45
Sodium phosphate (dibasic)[a]	162
Sodium phosphate (tribasic)[a]	54
Sodium tripolyphosphate[a,b]	900
Tetrapotassium pyrophosphate[a,b]	47.7
Tetrasodium pyrophosphate[a]	100
Triple superphosphate[a]	392

Source: *Chemical Economics Handbook*, 1969.
[a] Not P_2O_5 equivalent, but actual product.
[b] Source: Athanassiadis, 1969.

cedures. . . . It is hoped that some of these discrepancies may be resolved at a later date.

It is noted that the Bureau of Mines has published statistics on rock distribution based on information provided by producers.* These data are considered suspect by some industrial

* For 1966, the *Minerals Yearbook* (1967) figures for various categories in thousands of metric tons of P_2O_5 equivalent were wet phosphoric acid and

sources and therefore have not been used in the main consumption table of this report. However, it is still quite possible that they are the most accurate data available.

The Stanford Research Institute (1967), in its publication *Chemical Origins and Markets,* has developed a chart showing the names and end uses of products derived from the processing of phosphate rock. Due to space limitations, we are unable to reproduce the whole chart, but the names of many of the products are included in Table 5.11. Production figures for some fertilizers containing phosphorus (in thousands of metric tons of P_2O_5 equivalent) are direct phosphate rock application, 10.8; from furnace phosporic acid, 220; and from wet process phosphoric acid, 3,180 (*Chemical Economics Handbook,* 1969).

Data on the production of many of the derivative compounds can be found in the *Chemical Economics Handbook* and are reproduced in Table 5.12.

Table 5.11 and Table 5.12 reveal some of the difficulties that will be encountered in developing accurate determinations of the amount of phosphorus that enters the environment throughout the United States and the world. Such information will have to be available, however, if man is to control effectively the release of this potentially serious pollutant into lakes, rivers, and estuaries.

References

American Petroleum Institute (API), 1967. *Petroleum Facts and Figures* (New York: API).

American Petroleum Institute (API), 1968. *1967 Domestic Refiney Profile* (New York: API).

American Petroleum Institute (API), 1969. *Offshore Petroleum and the Environment* (New York: Committee on Public Affairs, API).

American Petroleum Institute (API), 1970a. *Final Report of the Task Force on Used Oil Disposal* (New York: API).

American Petroleum Institute (API), 1970b. *Systems Study of Oil Spill Cleanup Procedures* (New York: API).

Athanassiadis, Y. C., 1969. *Air Pollution Aspects of Phosphorus and its Compounds* (Bethesda, Maryland: Litton Systems, Inc.).

Backman, W. A., 1969. Forecast for the 70's, *Oil and Gas Journal.*

Chemical Economics Handbook, 1969. (Menlo Park, California: Stanford Research Institute).

Federal Water Pollution Control Administration (FWPCA), 1968. *Cost of Clean Water,* Vol. III (Washington D.C.: U.S. Government Printing Office).

triple superphosphate, 4,886; normal superphosphate, 1,793; elemental phosphorus, 1,678; other, 198; exports, 2,803; and total, 11,358.

Food and Agriculture Organization (FAO), 1969. *Yearbook on Food and Agricultural Statistics* (Rome: FAO).

Glazier, F. P., and Sumner, N.E., 1970. Decreasing relative demand for lubricants (San Antonio: National Petroleum Refiners Association Meeting), Report #AM–70–21.

Inter-governmental Maritime Consultative Organization, 1965. Report OP/n6. I/3.

McGraw, M. *Air Pollutant Emission Factors* (National Air Pollution Control Administration), forthcoming.

Marine Resources and Legislative and Political Arrangements for their Development, 1969. Panel Reports for the Commission on Marine Science, Engineering, and Resources, Vol. III.

Minerals Yearbook, 1967. (Washington, D.C.: U.S. Bureau of Mines, U.S. Government Printing Office), 1968.

Minerals Yearbook, 1968. (Washington, D.C.: U.S. Bureau of Mines, U.S. Government Printing Office), 1969.

Moss, J. E., 1963. *Oil Pollution in the Sea* (New York: American Petroleum Institute).

National Air Pollution Control Administration (NAPCA), 1970.

Nationwide Inventory of Air Pollutant Emissions (Raleigh, North Carolina: NAPCA).

Oil and Gas Journal, 1968. Supertanker size to mushroom.

Revelle, R., 1963. Water, *Scientific American, 209.*

Stanford Research Institute (SRI), 1967. *Chemical Origins and Markets* (Menlo Park, California: SRI).

Sullivan, R. J., 1969. *Air Pollution Aspects of Iron and Its Compounds* (Bethesda, Maryland: Litton Systems, Inc.).

United Nations. *Statistical Yearbook,* 1969 (New York: Statistical Office of the United Nations), 1970.

United Nations. *World Energy Supplies,* Statistical Papers, Series J. (New York: Statistical Office, Department of Economic and Social Affairs, United Nations).

U.S. Coast Guard, 1969. *Oil Spills in Pollution of U.S. Waters.*

U.S. Department of Agriculture (USDA), 1969. *Pesticide Review.*

U.S. Department of Health, Education, and Welfare, 1963. *Public Health Service Water Pollution Surveillance System Annual Compilation of Data, October 1, 1962—September 30, 1963* (Washington, D.C.: U.S. Government Printing Office).

U.S. Tariff Commission, 1970. *Synthetic Organic Chemicals, U.S. Production and Sales of Pesticides and Related Products* (Washington, D.C.: U.S. Government Printing Office).

6.
Work Group on Domestic and Agricultural Wastes

Chairman
Norman H. Brooks
CALIFORNIA INSTITUTE OF TECH-
NOLOGY

Geirmundur Arnason
CENTER FOR THE ENVIRONMENT
AND MAN, INC.

John F. Brown, Jr.
GENERAL ELECTRIC RESEARCH AND
DEVELOPMENT CENTER

M. Grant Gross
STATE UNIVERSITY OF NEW YORK

Bruce B. Hanshaw
U.S. GEOLOGICAL SURVEY

J. B. Hilmon
U.S. FOREST SERVICE

Philip C. Kearny
AGRICULTURAL RESEARCH SERVICE

George Rathjens
MASSACHUSETTS INSTITUTE OF
TECHNOLOGY

Rapporteur
Robert E. Stoller
HARVARD LAW SCHOOL

6.1
Introduction

Production on a global scale of certain airborne and water-borne materials was estimated for domestic, agricultural, and mining activities exclusive of materials discharged by industrial or energy-generating processes. These production figures assume steady-state production and do not deal with large, highly localized discharges of materials that cause many local and regional pollution problems. Because of the limited data base available no attempt was made to project these discharge figures into the future.

6.2
Sources of Materials Released to the Atmosphere

Natural production of airborne particles was found to be the largest source of these materials in the sectors considered. The available data come from two different approaches. The first involves estimates by the National Air Pollution Administration (NAPCA) of natural inputs from identifiable sources (Table 6.1). The second involves estimates based on the observed concentrations of mineral particles in glacial ice and the concentration of sea salts in rain (Table 6.2). The estimates based on the second approach indicated large fluxes (probably minimal estimates) but provided little information about probable sources of the materials. It was not possible, for example, to determine what fraction of the observed fallout of fine mineral grains resulted from wind erosion of deserts and high mountains that are little affected by man's activities, as opposed to wind erosion of agricultural lands influenced by man.

These data provide a sense of the order of magnitude of the movements of particles through the atmosphere. The precision of these estimates, however, is limited by uncertainties in observations of fallout rates as well as lack of available data on particle size distributions, which determine atmospheric residence times. The group did not consider production of particles formed through chemical reactions among gaseous atmospheric constituents.

For the period 1850–1970 Mitchell has calculated that the average stratospheric loading of very fine particulate matter (0.1–1 micron) due to volcanic activity was roughly 4.2 million metric

Table 6.1
Selected Sources of Airborne Particles Based on Inputs
(Million metric tons per year)

Sources	U.S. Total Particles of All Sizes[a]	U.S. 5-micron-diameter Particles	U.S./World Ratio	Estimated World Release 5-micron-diameter Particles[b]
Agriculture and forestry burning				
Forest	6.09	0.9[e]	6[d]	5.4
Crop wastes and ranges	2.18	0.33[e]	10[e]	3.3
Forest cleaning	(no reliable available estimates . . . .)			
Grain processing	0.73	0.02[f]	10[g]	0.2
Mining	1.45	(no reliable available estimates . . .)		
Waste incineration	1.0	0.18[h]	( )[i]	——

[a] NAPCA, 1970.

[b] Computed from columns 2 and 3 of this table.

[e] Duprey, 1968, gives 15 percent of all particles released by all types of refuse incinerators as less than 5 microns in diameter. In the absence of other data, we have taken 15 percent as the fraction of particles less than 5 microns emitted by burning.

[d] The United States has about one-sixth of the forest acreage under exploitation and also about one-sixth of the world forest area. Therefore, to convert U.S. data on forest fire emissions to a world estimate, we have multiplied by a factor of 6.

[e] The United States has about one-ninth of the cropland, pastureland, and rangeland of the world. We have added an additional factor to account for the greater use of fire in other continents.

[f] Duprey, 1968, cites one test of particle size distribution from feed and grain mills in which 3 percent of the particles by weight determination were less than 5 microns in diameter.

[g] Using the 1967 production of wheat flour, both global and U.S. (United Nations, *Statistical Yearbook*, 1968), the United States accounts for one-tenth of world production.

[h] NAPCA, 1969. The percent (by weight) of all released particles that are less than 5 microns in diameter at two municipal incinerators have been analyzed: Los Angeles 30 percent, and Milwaukee 6 percent. The figure in the table assumes the average of these two numbers, 18 percent for the entire United States. This assumption is not necessarily correct, but the results are illustrative.

[i] No estimate is available for this figure. It might be noted that in much of the industrialized world (Europe) disposal methods are in many cases better than those in the United States, while in the underdeveloped world waste incineration is small relative to other sources of particles and compared to that in industrialized countries.

Table 6.2
Sources of Airborne Particles Based on Observed Atmospheric Fallout
(Million metric tons/yr)

Sources	Observed Fallout
Sea salts	1,000 to 2,000[a]
Mineral dusts	
Northern Hemisphere	50[b]
Southern Hemisphere	20[c]

[a] Source: Woodcock, A. H., 1962. This estimate indicates the probable rate of production at the sea surface of salt particles that remain airborne long enough to be transported up to local cumulus-base altitudes.

[b] The rate of dust production of a given area is a poorly understood function of numerous variables including the size of its land surface, its topographical features (mountains, forest, desert), and the use to which it is put. Nevertheless, a minimum estimate of the annual production of atmospheric dust can be made by assuming steady state and using measurements (Windom, 1969) of dust accumulations in five North American glaciers between Greenland (77°N) and Mt. Popocatepetl, Mexico (19°N) in the three major wind systems of the Northern Hemisphere. Particles accumulate at a median rate of about 0.00002 g/cm^2-yr (ranging from 0.07 to 3.2×10^{-5} g/cm^2-yr). (No size data for particles were given.) Assuming that fallout was uniform over the Northern Hemisphere (2.55×10^{18} cm^2), the total fallout was about 5×10^{13} g/yr, or 50 million metric tons per year.

[c] Less dust is produced in the Southern Hemisphere owing to the relative scarcity of arid lands there as compared to the Northern Hemisphere (Bogdanov, 1963). We assumed that the dust production of the Southern Hemisphere was 40 percent of the Northern Hemisphere.

tons (Mitchell, forthcoming). The variability of rates from one year to the next is, or course, great. During the Krakatoa (1888) and Agung (1963) eruptions the output may have been respectively as much as 12 and 4 times greater than the average figure (Mitchell, forthcoming).

Mitchell assumes that for major eruptions 1 percent of the ejected matter reaches the stratosphere while the remaining 99 percent consists mainly of blocks and large particles that fall rapidly back to earth and do not reach the stratosphere. He bases his calculations on an unpublished estimate of the total ejected mass for each recorded eruption since 1855 (there have been 39). For the 1 percent assumed to reach the stratosphere, Mitchell assumes a residence time of 14 months (Mitchell, forthcoming). The group found no other data on which to estimate volcanic input of particles into the atmosphere.

Agricultural and forestry wastes, especially burning of forests and forest wastes (Table 6.1) appear to be major contributors to airborne particles. Much of this material would not have been included in the estimates of fine mineral dusts in Table 6.2. Changes in agricultural land and pesticide use will affect the discharge of particles water and atmosphere. Estimates of these changes were not made by this Work Group because we lacked firm data on which to base projections.

Mining did not appear to be an important source of airborne or waterborne wastes on a global scale in comparison to the natural movements of materials. Although mining and other activities are prolific producers of wastes on a local or regional scale, the large particles produced are thought to remain near the point of origin and therefore constitute local or regional rather than global problems.

6.3

Sources of Pollutants Released to Estuaries and Oceans

We exclude sources that are industrial or related to energy conversion. Natural inputs to the ocean (nutrients, sediment, oil seeps) are limited to waterborne routes. Estimates on the amount of "natural" riverborne sediment brought to the ocean range from 10 to 30 billion metric tons per year, 65 to 80 percent coming from Asia (Strakhov, 1967; Holeman, 1968). In most rivers the coarse-grained material is transported along the bottom and trapped in the estuary or on the delta at the river mouth, thus remaining near the point of injection to the ocean (Strakhov, 1967). The remainder of the riverborne sediment, normally the finer-grained fraction, moves approximately parallel to the coast (Gross, 1966; Meade, 1969) with little escaping to the deep ocean floor.

Dredged wastes coming from urban areas contain not only sediment but other wastes as well, including sewage solids, agricultural and industrial wastes deposited in local waterways. Customarily these dredged materials are dumped in coastal waters. New deep-draft vessels need much greater channel depths (20 meters) than is presently offered in most port areas, necessitating

Table 6.3
Global Land Use
(10^6 hectares)

	Total	Cropland	Pasture Range	Forest	Other
1950	13,509	1,230	2,187	4,024	6,068
1957	13,670	1,384	2,407	3,839	6,040
1968	13,395	1,447	2,892	3,994	5,062

Sources: Food and Agriculture Or-   *Yearbook on Food and Agricultural Sta-*
ganization (FAO) 1951, 1958, 1969.   *tistics,* 1950, 1957, 1967.

a substantial increase in dredging and in the volume of dredged wastes to be dumped offshore (Gross, 1970a).

Surveys of dredged wastes and barged waste disposal indicate that approximately 73 million metric tons per year of solids are dumped each year in the marine waters of the United States and Canada (Gross, 1970a, 1970b), increasing at an estimated 4 percent per year. (Large amounts of dredged wastes and industrial waste solids are also dumped each year into the Great Lakes.) This 73 million metric tons is approximately 4 percent of the natural sediment deposit from erosion (Holeman, 1968). If the amount of dredging is related to the port activity, it seems reasonable to assume that world dredged waste discharge must be at least twice to three times that of the United States.* Thus, the estimated world discharge of dredged wastes is probably at least 150 to 220 million metric tons per year.

6.4

Agricultural Wastes and Practices

This section deals with global land use, pesticide use, and fertilizer production and consumption. Data on global land use are shown in Table 6.3.

A second area of concern is pesticide use. Pesticides are interpreted here to mean all classes of biocides including insecticides, fungicides, nematocides, and others. For agricultural pur-

---

* Estimations of the worldwide discharge of dredged wastes were made by noting that 6 of the 10 largest ports are in North America and handled 57 percent of their total tonnage in 1967.

Table 6.4
Toxicities and Hazards of Some Insecticides That Will Be Substituted for DDT

Insecticide	Acute Oral LD$_{50}$[a] Rates (mg/kg)	Acute Dermal LD$_{50}$[b] Rates (mg/kg)
Phorate	1.1– 2.3	2.5– 6.0
Demeton	2.5– 6.2	8 – 14
Parathion	3.6– 13.0	7 – 21
Ethion	27 – 65	62 –245
DDT	113 –118	2,510

Source: *Farm Chemicals*, 1970.
[a] Oral intake that has lethal effects on 50 percent of a test population.
[b] Contact with the skin that has lethal effects on 50 percent of a test population.

poses the more-developed countries use the most pesticides (see Table 2.6 in the report of Work Group 2).

Significant changes in pesticide use are likely in the future. In the United States it is likely that there will be a further decrease in the use of the persistent organochlorine pesticides (DDT, dieldrin, toxaphene, heptachlor chlorden, endrin) and significant increases in the use of insecticides such as organophosphorus (phorate, demeton, parathion, and diazion) and methyl carbamate (sevin, baygon) insecticides. The toxicity of these latter pesticides, which have been proposed as substitutes, are higher than that of DDT (Table 6.4), but their persistence is less. Therefore, in order to obtain the same degree of insect control as previously achieved with DDT, several applications of the less-persistent insecticides must be sprayed on the same area.

Markets outside the United States will probably continue to increase their use of organochlorine and organophosphorus insecticides. Estimates of the projected world use of organochlorine pesticides will depend on the ability of developing nations to import or manufacture these pesticides and the bans imposed in the more developed countries on the use of these same pesticides. Estimates will also depend on the goals in food production established in each of the developing countries. The total increase in pesticides required to meet the needs of the major developing

Table 6.5
Pesticides Needed to Increase Food Production on Acreage Now under Cultivation in Asia (Except Mainland China and Japan), Africa, and Latin America by the Percentages Indicated

Percentage of Increase in Agricultural Production	Tonnage Needed (metric tons)
⸺	120,000
10	150,000
20	195,000
30	240,000
40	285,000
50	342,000
60	402,000
70	475,000
80	558,000
90	640,000
100	720,000

Source: President's Science Advisory Committee (PSAC), 1967.

areas (Asia, Africa, and Latin America) for the designated increased productions levels are given in Table 6.5.

Finally we considered fertilizer production and consumption. The three major nutrients in chemical fertilizers are nitrogen (N), phosphorus (P, usually in P_2O_5), and potassium (K, usually in K_2O). Except for the years 1914–1918 and 1940–1945, world production and consumption has doubled or tripled in each decade. The world consumption of fertilizer nutrients and compound rate of increase for the decade 1954 to 1964 has been calculated by the FAO (see Table 6.6).

Although very little fertilizer is now used in Africa, Asia, and Latin America, fertilizer consumption will increase greatly in developing countries in the next decade. Table 6.7 shows the higher rate of fertilizer production and consumption in developing, as compared to developed, countries.

Nevertheless, problems associated with large-scale fertilizer production and usage will be concentrated in the developed countries for at least the next decade. The total world use of N, P, and K in 1963–1964 exceeded 33 million metric tons, with only about 3.3 million metric tons used in developing countries (Table 6.6).

Table 6.6
**World Consumption of Fertilizer Nutrients and Compound Rate
of Increase, 1953/1954–1963/1964**
(Thousands of metric tons)

Area	1954	1964	Compound Rate of increase (percent)	
			1954–1964	1959–1964
Developed regions				
Western Europe	6,450	11,045	5.5	5.6
Eastern Europe	2,789	6,327	8.5	7.7
North America	5,280	9,101	5.5	7.1
Oceania	560	1,117	7.1	10.0
Other[a]	1,037	2,123	7.4	4.2
Total	16,116	29,713	7.0	6.5
Developing regions				
Africa	104	353	13.0	12.3
Asia[b]	445	1,840	15.2	16.6
Latin America	416	1,167	10.8	12.2
Total	965	3,360	11.6	14.5
World Total	17,081	33,073	7.4	7.2

Source: Food and Agriculture Organization of the United Nations (FAO), *Fertilizers: An Annual Review of World Production, Consumption and Trade, 1954–1964*, 1965.

[a] Developed countries within developing regions: Japan, South Africa, and United Arab Republic.
[b] Excluding Mainland China.

Even in 1975–1976, production in the developing countries will represent only 14 percent of the world production of N, 10 percent of the world production of P, and 5 percent of the world production of potash (K_2O) (UNIDO, 1969).

Recommendations
We recommend that data be gathered in the following areas:
1.   The effect of changed land use on the production of airborne particles. This seems to be especially important in arid and semiarid areas with fine-grained soils that were originally windblown deposits and therefore susceptible to wind erosion.
2.   Aspects of mining, ore handling, and bulk processing of

Table 6.7
**Estimated Production and Consumption of Fertilizers
by Region, 1970/1971 and 1975/1976**
(Thousand metric tons of nutrients)

	Developed Areas	Developing Areas	World[a] Total
Projection for 1970/1971			
Production	63,300	5,800	69,100
Consumption	56,000	9,500	65,500
Surplus/(deficit)	7,300	(3,700)	3,600
Projection for 1975/1976			
Production	88,500	10,700	99,200
Consumption	79,000	16,150	95,150
Surplus/(deficit)	9,500	(5,450)	4,050

Source: United Nations Industrial Development Organization, 1969.

[a] Excluding China (mainland), North Korea, and North Viet-Nam.

known toxic materials should be investigated to permit better assessments of their potential (or existing) environmental impact. The group was unable to obtain reliable data on waste discharges from this potentially important sector. Detailed information is needed on bulk- and elemental-emission rates.

References

Bogdanov, D. V., 1963. Map of the natural zones of the ocean, *Deep-Sea Research, 10.*

Duprey, R. L., 1968. *Compilation of Air Pollution Emission Factors* (Raleigh, North Carolina: United States Department of Health, Education, and Welfare).

Farm Chemicals, January 1970, *133.*

Food and Agriculture Organization (FAO), 1951, 1958, 1969. *Yearbook on Food and Agricultural Statistics,* 1950, 1957, 1968 (Rome: FAO).

Food and Agriculture Organization (FAO), 1965. *Fertilizers: An Annual Review of World Production, Consumption, and Trade, 1954–1964* (Rome: FAO).

Food and Agriculture Organization (FAO), 1963. *Production Yearbook* (Rome: FAO).

Gross, M. G., 1966. Distribution of radioactive marine sediment from the Columbia River, *Journal of Geophysical Research, 71.*

Gross, M. G., 1970a. Waste-solid disposal in coastal waters of North America.

Gross, M. G., 1970b. New York Metropolitan Region—a major sediment source, *Water Resources Research, 6.*

Holeman, J. N., 1968. The sediment yield of major rivers of the world, *Water Resources Research, 4.*

Meade, R. H., 1969. Landward transport of bottom sediment in estuaries of the Atlantic coastal plane, *Journal of Sedimentary Petrology, 39.*

Mitchell, J. M., Jr., 1970. A preliminary evaluation of atmospheric pollution as a cause of the global temperature fluctuation of the past century, *Global Effects of Environmental Pollution,* edited by J. F. Singer (Dordrecht: Reidel Publishing Company), forthcoming.

National Air Pollution Control Administration (NAPCA), 1969. *Air Quality Criteria for Particulate Matter* (Washington, D.C.: NAPCA).

National Air Pollution Control Administration (NAPCA), 1970. *Nationwide Inventory of Air Pollutant Emissions* (Raleigh, North Carolina: NAPCA).

President's Science Advisory Committee (PSAC), 1967. *The World Food Problem* (Washington, D.C.: U.S. Government Printing Office).

Strakhov, N. M., 1966. *Principles of Lithogensis,* Vol. I (New York: Consultants Bureau Enterprises, Plenum Publishing Corp.).

United Nations. *Statistical Yearbook,* 1968 (New York: Statistical Office of the United Nations), 1969.

United Nations Industrial Development Organization (UNIDO), 1969. Fertilizer Industry, UNIDO Monograph on Industrial Development No. 6 (New York: United Nations).

Windom, H. L., 1969. Atmospheric dust records in permanent snow fields: implications to marine sedimentation, *Bulletin of the Geological Society of America, 80.*

Woodcock, A. H., 1962. Solubles, *The Sea,* Vol. I, edited by M. N. Hill (New York: Interscience [Wiley]).

7.
Work Group on Energy Products

Chairman
Edward Hamilton
THE BROOKINGS INSTITUTION

John Franklin
GREENWICH, CONNECTICUT

Allan V. Kneese
RESOURCES FOR THE FUTURE, INC.

Frank G. Lowman
PUERTO RICO NUCLEAR CENTER

George B. Morgan
NATIONAL AIR POLLUTION CON-
TROL ADMINISTRATION

Jerry S. Olson
OAK RIDGE NATIONAL LABORA-
TORY

Silvio Simplicio
WEATHER BUREAU

Rita F. Taubenfeld
SOUTHERN METHODIST UNIVER-
SITY

Herbert L. Volchok
ATOMIC ENERGY COMMISSION

Rapporteur
Terry L. Schaich
FLETCHER SCHOOL OF LAW AND
DIPLOMACY

7.1

Introduction

We desired, to the extent possible in a week of effort at some distance from many sources of data, to examine existing projections for responses to four principal questions: (1) What is the likely range of energy consumption in the United States and in the world (a) at present and (b) in 1980 and 2000? (2) What is the volume of pollutants likely to be generated, given various assumptions about how the necessary energy will be produced? (3) How is the production of these pollutants likely to be distributed around the globe? (4) What is the potential effectiveness of known pollution-control technology to reduce the pollution coefficients of various forms of energy production, and what order of magnitude of difference could full application of these techniques produce in the global projections?

We discovered that the present state of knowledge allows a detailed response to only the first question, and that even that response is necessarily speculative and error-prone. In this report we present the best data we could find with respect to each of the four questions.

7.2.

Energy Consumption

Looking at the United States first, we found conflicting figures for the total U.S. consumption of energy even for 1968, the latest year for which firm data are widely available. The discrepancies are of the order of 10 percent of the total figure and result primarily from two factors: (1) Many (but not all) sources of data add nonenergy uses of fuel to their totals; and (2) varying rates are used to convert hydroelectricity into units such as Btu's or kilowatt-hours thermal. Both of these factors were discussed in the *Review and Comparison of Selected United States Energy Forecasts* developed in 1969 by Battelle Memorial Institute.*

A useful set of figures for energy consumption in the United States in 1968 in presented in Table 7.1. The advice of the

* This study is by far the best source for our purposes with regard to estimates and projections of U.S. energy consumption. For a fuller citation, see Battelle, 1969, in the references at the end of the report.

Table 7.1
U.S. Energy Consumption, 1968[a]
(10^{12} kilowatt hours, thermal)

Fuel Sector	Util-ities	In-dustry	Trans-port	Household, Commer-cial, Misc.	Total
Solid					
Bituminous coal and lignite	2.07	1.58	0.00	0.13	3.78
Anthracite coal	0.02	0.02	——	0.04	0.08
Total solid fuels	2.09	1.60	0.00	0.17	3.86
Liquid					
Liquified gases	——	0.02	0.04	0.21	0.27
Jet fuel: naphtha	——	——	0.20	——	0.20
Jet fuel: kerosene	——	——	0.37	——	0.37
Gasoline	——	——	3.01	——	3.01
Kerosene	——	0.04	——	0.13	0.17
Distillate fuel oil	0.01	0.11	0.36	1.00	1.48
Residual fuel oil	0.34	0.31	0.23	0.36	1.24
Still gas	——	0.26	——	——	0.26
Petroleum coke	——	0.10	——	——	0.10
Total liquid fuels	0.35	0.84	4.21	1.70	7.10
Natural gas, dry	0.95	2.58	0.18	1.89	5.60
Hydropower[b]	0.22	——	——	——	0.22
Nuclear power[b]	0.04	——	——	——	0.04
Total	3.65	5.02	4.39	3.75	16.82

Source: *Minerals Yearbook*, 1968, Review of minerals industry, Tables 7, 9, 11, 12, 13, and 14.

[a] *Minerals Yearbook* Btu figures were converted into kWh (thermal) at 3,412 Btu/kWh(t) or 0.293×10^{-3} kWh(t)/Btu.

[b] Figures on generation of electricity from hydro and nuclear sources were taken from Table 7 of the overall review section of the 1968 *Minerals Yearbook*; kWh(e) were converted to kWh(t) as follows:

(1) For hydro, kWh(t)

$$= \left[\frac{3{,}412 \text{ Btu}}{\text{kWh(e)}} \frac{\text{kWh(t)}}{3{,}412 \text{ Btu}} \right] \text{kWh(e)}$$

$$= [1] \text{ kWh(e)}$$

(2) For nuclear, the *Minerals Yearbook* figure for average central station conversion rates (i.e., 10,582 Btu/kWh(e) for 1968) was used to convert to Btu's which were then converted to kWh(t) at 0.293×10^{-3} kWh(t)/Btu.

Battelle study was followed with respect to the preceding two factors. Hence, nonenergy uses of fuels were excluded and electricity from hydropower was evaluated by including only the energy contained in the generated electricity and not the additional energy that would be required to generate the same amount of electricity in a central thermal power plant.

The conflicting definitions and terminologies used in publications on energy and fuel consumption make it difficult to deal with projections of these figures. The Battelle report notes:

> With these considerations in mind, it is evident that direct comparisons among energy forecasts by various investigators ran be misleading. Direct comparisons are strictly valid only for those studies using equivalent definitions. Theoretically, one could "correct" the forecast values by the various investigators to some desirable common definition. However, in most cases this is almost impossible to accomplish since the reports do not present the necessary information, data, and references which would allow such a correction to be made.

Nonetheless, the same authors conclude:

> General trend and order of magnitude comparisons can be informative even for two studies based on slightly different definitions—especially so if the definition differences are kept in mind while making the comparison (Battelle, 1969).

In this spirit, we present in Table 7.2 a simple summary of forecasts taken from the Battelle study.

Using the data in that table and in the many others in the Battelle report, the Energy Policy Staff of the U.S. Office of Science and Technology noted in the foreword to the Battelle study:

> According to the forecasts examined in this report, energy consumption in the year 2000, including non-fuel uses, is expected to be about 170,000 trillion British thermal units [50 trillion kilowatt-hours thermal] if real gross national product grows at about 4 percent per year. Consumption in 1968 [including non-fuel uses and using a central plant conversion rate factor to evaluate hydro electricity] was slightly over 62,000 trillion BTU [over 18 trillion kilowatt-hours thermal]. The average annual indicated growth rate is about 3.2 percent (Battelle, 1969).

The Energy Policy Staff also pointed out the many weak points in the foundation underlying these projected figures. Their comments are an essential caveat to anyone who would use these projections:

> Review of the forecasts [included in the Battelle study] . . . reveals that many relevant questions are not answered or in many cases even addressed. Before it is possible to gain a high degree of confidence in even the range within which future energy consumption is likely to fall, many of these questions must be considered and answered.

Table 7.2
Forecasts of Total Energy Requirements for United States Only
(10^{12} kilowatt-hours, thermal)

Source Document[a]	Date of Publication	Growth Rate[b] Base Year to 1980 (percent)	1980	Growth Rate[b] 1980 to 2000 (percent)	2000
CGAEM[c]	1968	3.7	28.6	——	——
EUS[c,e]	Sept. 1967	4.2	27.4	——	——
OEUS	Oct. 1968	3.8	28.4[d]	——	——
USP	July 1968	3.3	25.8	——	——
EMUS	July 1968	3.2	25.8	3.2	49.4
		3.2	24.6[e]	3.2	46.5[e]
PCCP[e]	May 1968	3.5	26.7	3.1	45.4
TCUSEC[f]	1968	3.2	26.5	3.3	51.0
		(3.7)	(29.2)	(3.9)	(62.5)

Source: Battelle, 1969. Converted from Btu's at 0.293×10^{-3} kWh(t)/ Btu.

[a] Source documents are as follows:

CGAEM:
Competition and Growth in American Energy Markets, 1947–1958. Texas Eastern Transmission Corporation.

EUS:
Energy in the United States, 1960–1985. Michael C. Cook, Sartorius & Co.

OEUS:
Outlook for Energy in the United States. Energy Division, The Chase Manhattan Bank, N.A.

USP:
United States Petroleum Through 1980. United States Department of Interior, Office of Oil and Gas.

EMUS:
An Energy Model for the United States, Featuring Energy Balances for the Years 1947–1965 and Projections to the years 1890 and 2000. Bureau of Mines, IC8384, United States Department of Interior.

PCCP:
Projections of the Consumption of Commodities Producible on the Public Lands of the United States 1980–2000. Prepared for the Public Land Law Review Commission by Robert R. Nathan Associates, Inc., Washington, D.C.

TCUSEC:
Technological Change and United States Energy Consumption, 1939–1954. Alan M. Strout (unpublished thesis) University of Chicago.

[b] Growth rates generally indicate the compound annual rate of growth from the average value for one period to the average of another period.

[e] Hydro accounted for at kWh energy equivalent.

[d] Converting their 17,000 million barrels of oil equivalent to Btu at 5,800,000 Btu per barrel.

[e] Excludes nonfuel uses.

[f] GNP growth rate at 3.5 percent per year and (4.0 percent peryear).

On the level of total energy consumption, there has been a definite relationship between total energy consumption and real gross national product in the past. Over the last 50 years the relationship shows that a decreasing amount of energy has been required for each unit of GNP. The increased technical efficiency of energy use has tended to more than offset the more intense use of energy in our economy. However, the trend appears to be changing. In the future, it is possible that a constant or even increasing amount of energy per unit of GNP may be required if present policies of encouraging energy use are continued. One reason for the changing trend is that the technical efficiency of new electric power plants and many other energy conversion devices is no longer increasing and may even decrease over the next several decades. This factor, coupled with the increasing share of end uses being supplied by electricity, is at least an important item working in the direction of changing the historical relationship. It is thus possible that most projections have understated future growth in the overall energy consumption if present trends continue.

The forecasts reviewed [in the Battelle study] were prepared before the recent surge of concern about the environment. They contain little information about the effects of environmental quality control on energy consumption even though the production, transportation, and utilization of energy is deeply involved with environmental quality and conservation considerations. It is possible that in the future government policy in regard to environmental quality and conservation matters will lead to a lower level of total energy consumption than would otherwise occur. In fact, if environmental quality control considerations and other factors increase the cost of energy over what would otherwise prevail, that fact alone will to some extent decrease consumption, the extent of the decrease depending, of course, on the magnitude of cost increases and the elasticity of energy demand. . . .

Widespread use of fuel cells or heat pumps could tend to increase technical efficiency. Similarly, various structural changes in the economy, such as the widespread displacement of private automotive travel by mass transportation, might act in the direction of decreasing energy consumption per unit of GNP.

In the case of individual fuels, most of the forecasts made several convenient but questionable assumptions. A major assumption is that overall and relative prices for energy will be such that price need not be explicitly considered in the forecasts. A second and related assumption is that there will be no limit on the availability of any fuel or energy form. The assumption of unlimited availability at no change in relative prices is of questionable validity for even the next decade, let alone the remainder of the century. It is, of course, very difficult to consider price and resource availability in energy forecasts, but it appears to be increasingly necessary to do so (Battelle, 1969). In short, the preceding figures *may* be a rough indication of the scale of energy needs and by-products for the United States alone if future developments follow previous trends. They are certainly no more than that, and may very well be much less.

With respect to worldwide consumption of energy, the

historical data exhibit the same sort of conflicts in coverage and methods of evaluation that were evident in the figures for the United States alone. The problems are not immediately apparent, since for 1967 (the last year for which data are available) the U.N. figures for overall U.S. energy consumption agree very well with figures for 1967 computed from data in the *Minerals Yearbook* for 1968 by the method used in Table 7.1. However, the data and assumptions used for these two sets of figures differ in several respects. Table 7.3 contains the world energy consumption data compiled by the United Nations; as noted, however, the figures for electricity from nuclear sources have been calculated using the Battelle system referred to earlier.

There are very few worldwide energy consumption projections. One is cited in section 1.2.3 of the report of Work Group 1. Another that extends to 1980 was done by Joel Darmstadter of Resources for the Future in a work entitled "Energy and the World Economy" (to be published by The Johns Hopkins Press).* Some salient figures from this projection are presented in Table 7.4.

7.3

Emissions from Energy Consumption

Published data on emissions from power production and from other forms of energy consumption are not firm for the United States and are virtually nonexistent for the world as a whole.

There are, however, some emissions that are very easy to calculate, once one knows the total consumption of fuel in the world. In the case of waste heat, for example, the total energy consumed can be considered to be heat released to the environment. Or if it is thought desirable to exclude the heat that results from the use of electricity since that heat may take longer to reach the environment than the heat from a furnace, for example, then all that need be done is to subtract the heat con-

* Another projection has been made by Chairman Seaborg of the U.S. Atomic Energy Commission before a Congressional committee:

Assuming world use [of all forms of energy] at one-quarter of the U.S. per capita consumption, this would, in the year 2000, amount to 650 quadrillion British thermal units [190 trillion kilowatt-hours thermal] of energy needed for 6 or 7 billion people, a staggering total. (Seaborg, 1969, p. 86.)

Table 7.3
World Energy Consumption, 1967[a]

	Solid		Liquid		Gas		Hydro[b]		Nuclear[c]		Overall	
	10^{12} kWh(t)	Percentage of World Consumption	10^{12} kWh(t)	Percentage of World Consumption	10^{12} kWh(t)	Percentage of World Consumption	10^{12} kWh(t)	Percentage of World Consumption	10^{12} kWh(t)	Percentage of World Consumption	10^{12} kWh(t)	Percentage of World Consumption
Developed Countries												
United States	3.52	20.2	6.33	35.7	5.58	64.1	0.22	21.8	0.03	25.	15.68	34.8
Canada	0.18	1.0	0.63	3.6	0.38	4.4	0.13	12.9	—	—	1.32	2.9
Western Europe	3.67	21.1	4.42	24.9	0.33	3.8	0.32	31.7	0.09	75.	8.83	19.6
Eastern Europe	2.42	13.9	0.37	2.1	0.26	3.0	0.01	1.0	—	—	3.06	6.8
USSR	3.47	19.9	2.23	12.6	1.66	19.1	0.09	8.9	n.a.[d]	n.a.	7.45	16.6
Japan	0.61	3.5	1.11	6.3	0.02	0.2	0.07	6.9	—	—	1.81	4.0
Oceania	0.26	1.5	0.24	1.4	—	—	0.02	2.0	—	—	0.52	1.2
Total	14.13	81.1	15.33	86.6	8.23	94.6	0.86	85.2	0.12	100.	38.67	85.9
Developing Countries												
Communist Asia	1.97	11.3	0.13	0.7	0.12	1.4	0.04	4.0	—	—	2.14	4.8
Other Asia (exc. Japan)	0.76	4.4	0.85	4.8	0.02	0.2	0.03	3.0	—	—	1.76	3.9
Africa	0.43	2.5	0.29	1.6			0.02	2.0	—	—	0.76	1.7
Other America	0.10	0.6	1.14	6.4	0.33	3.8	0.06	5.9	—	—	1.63	3.6
Total	3.26	18.7	2.41	13.5	0.47	5.4	0.15	14.9	—	—	6.29	14.0
World Total	17.39	100.[e]	17.74	100.[e]	8.70	100.	1.01	100.[e]	0.12	100.	44.96	100.[e]

Source: United Nations, World Energy Supplies.

[a] U.N. figures are in million metric tons of coal equivalent. The energy contained within 1,000 generated kWh(e) of electricity is equated by the U.N. to 0.125 metric ton coal equivalent. Since the energy content of one kWh(e) = the theoretical 3,412 Btu, therefore 3,412 × 10^3 Btu = 0.125 m.t.c.e. and 1.0 m.t.c.e. = 27.3 × 10^6 Btu.

[b] Hydroelectricity is converted from kWh(e) production figures in Table 13 of United Nations, World Energy Supplies by the same method used for hydro in Table 7.1.

[c] Nuclear electricity is handled as in Table 7.1 using the same central station conversion rate.
Note: Nuclear totals exclude USSR, for which separate data are not available.

[d] Not available.

[e] Does not add to total due to rounding.

Table 7.4
Darmstadter's Projection of World Energy Consumption in 1980

	A[a]	Solid 10¹² kWh(t)[c]	Solid Percentage of World Consumption	Liquid 10¹² kWh(t)[c]	Liquid Percentage of World Consumption	Gas 10¹² kWh(t)[c]	Gas Percentage of World Consumption	Hydro[b] 10¹² kWh(t)[c]	Hydro[b] Percentage of World Consumption	Nuclear[b] 10¹² kWh(t)[c]	Nuclear[b] Percentage of World Consumption	Overall 10¹² kWh(t)[c]	Overall Percentage of World Consumption
Developed Countries													
United States	3.5	5.0	17.3	9.4	25.3	8.3	41.9	0.34	18.1	0.98	52.0	24.0	26.8
Canada	5.5	0.3	0.9	1.2	3.2	1.0	4.8	0.22	11.6	0.05	2.8	2.8	3.0
Western Europe	4.0	2.7	9.4	9.2	24.9	2.1	10.3	0.46	24.1	0.63	33.5	15.1	16.8
Communist Eastern													
Europe	4.6	3.6	12.5	1.5	3.9	0.7	3.5	0.02	1.2	0.04	2.1	5.9	6.6
USSR	6.5	5.7	19.7	5.2	14.0	5.9	29.8	0.29	15.3	0.04	2.1	17.1	19.1
Japan	7.9	0.5	1.9	3.5	9.5	0.1	0.4	0.11	5.7	0.11	5.6	4.3	4.9
Oceania	4.8	0.4	1.3	0.4	1.2	0.1	0.7	0.04	2.1	0.01	0.3	1.0	1.1
Total	4.7	18.2	63.0	30.4	82.0	18.2	91.4	1.48	78.1	1.86	98.4	70.2	78.3
Developing Countries													
Communist Asia	7.6	7.3	25.4	0.7	2.0	—	—[d]	0.04	2.2	0.01	0.4	8.1	9.1
Other Asia (exc. Jap.)	8.5	2.3	3.1	2.4	6.5	0.4	2.2	0.14	7.3	0.02	1.0	5.2	5.8
Africa	6.5	0.9	0.6	0.7	1.9	0.2	0.9	0.06	3.0	—	—	1.8	2.0
Other America	7.4	0.2	7.9	2.8	7.6	1.1	5.5	0.18	9.4	0.00	0.2	4.3	4.8
Total	7.7	10.7	37.0	6.6	18.0	1.7	8.6	0.42	21.9	0.03	1.6	19.4	21.7
World Total	5.2	28.9	100.0	37.0	100.0	19.9	100.0	1.90	100.0	1.89	100.0	89.6	100.0

Source: Estimated by Joel Darmstadter in *Energy and the World Economy* (to be published by The Johns Hopkins Press for Resources for the Future, Inc.).
[a] Column A contains the projected average annual percentage of growth in energy consumption for 1965–1980.
[b] Darmstadter follows the U.N. system of evaluating hydro and nuclear electricity. This means that he used for *both* nuclear and hydropower the system used by the Group *only* for hydropower in Tables 7.1 and 7.3. Darmstadter's actual figures were in metric tons of coal equivalent and were converted to kWh (both thermal and electrical in this case) at the U.N. rate of 1,000 kWh per 0.125 m.t.c.e. (for the factor see U.N., *World Energy Supplies* or the Appendix of any recent U.N. *Statistical Yearbook*).
[c] Converted from metric tons coal equivalent by using 27.3×10^6 Btu/m.t.c.e. and 0.293×10^{-3} kWh(t)/Btu.
[d] Unknown, but believed to be small.

Table 7.5

Nationwide Emissions in the United States, 1968[a]

(Million short tons per year)[b]

| | Emission | | | |
| | Particles[c] | | SO$_x$ | |
Source	10^6 short tons	Percentage	10^6 short tons	Percentage
Fuel Combustion	10.1	35.7	25.2	75.9
Transportation	1.2	4.3	0.8	2.4
Motor vehicles	0.8	2.8	0.3	0.9
Gasoline	0.5	1.8	0.2	0.6
Diesel	0.3	1.0	0.1	0.3
Aircraft[d]	N[e]	N	N	N
Railroads	0.2	0.7	0.1	0.3
Vessels	0.1	0.4	0.3	0.9
Nonhighway motor fuels use	0.1	0.4	0.1	0.3
Stationary sources	8.9	31.4	24.4	73.5
Coal	8.2	29.0	20.1	60.5
Fuel oil	0.3	1.0	4.3	13.0
Natural gas	0.2	0.7	N	N
Wood	0.2	0.7	N	N
Other Sources	18.2	64.3	8.0	24.1
Industrial processes[f]	7.5	26.5	7.3	22.0
Solid waste disposal	1.1	3.9	0.1	0.3
Miscellaneous[g]	9.6	33.9	0.6	1.8
Total	28.3	100.0	33.2	100.0

[a] Source: NAPCA, 1970. The authors of this document comment that the "numbers presented here should be representative of current emissions" but that "because of the increasing availability of more comprehensive data and emission factors, revisions will be made in emission quantities. Therefore, the numbers presented in this document are subject to change."

[b] Data were given in short tons and were not converted in the table to metric tons because of the distortion in the percentage figures that would result from rounding off the numbers in the emissions columns. Some of the overall figures are presented below in million metric tons:

Source	Particles	Oxides of Sulfur	Hydrocarbons	Oxides of Nitrogen	Carbon Monoxide
Fuels	9.2	22.9	15.7	16.5	59.7
Other	16.5	7.3	13.4	2.3	31.3
Total	25.7	30.2	29.1	18.8	91.0

tained in generated electricity (some 3.9 trillion kilowatt-hours thermal in 1967, according to data in the most recent issue of the United Nations' *World Energy Supplies* from total energy consumption as listed in Table 7.3.

In the case of carbon dioxide, one need only know the total weight of each fuel burned throughout the world and the percentage of carbon in each fuel. In using such a procedure one is assuming perfect combustion, and this perhaps would tend to raise the figures slightly, but the effect is probably small. A calculation of CO_2 emissions is detailed in the Appendix to this

		Emission			
HC		NOx		CO	
10^6 short tons	Percentage	10^6 short tons	Percentage	10^6 short tons	Percentage
17.3	54.1	18.1	87.8	65.7	65.7
16.6	51.9	8.1	39.3	63.8	63.8
15.6	48.8	7.2	34.9	59.2	59.2
15.2	47.5	6.6	32.0	59.0	59.0
0.4	1.3	0.6	2.9	0.2	0.2
0.3	0.9	N	N	2.4	2.4
0.3	0.9	0.4	1.9	0.1	0.1
0.1	0.3	0.2	1.0	0.3	0.3
0.3	1.0	0.3	1.5	1.8	1.8
0.7	2.2	10.0	48.5	1.9	1.9
0.2	0.6	4.0	19.4	0.8	0.8
0.1	0.3	1.0	4.8	0.1	0.1
N	N	4.8	23.3	N	N
0.4	1.3	0.2	1.0	1.0	1.0
14.7	45.9	2.5	12.2	34.4	34.3
4.6	14.4	0.2	1.0	9.7	9.6
1.6	5.0	0.6	2.9	7.8	7.8
8.5	26.5	1.7	8.3	16.9	16.9
32.0	100.0	20.6	100.0	100.1	100.0

e No estimates were given of the likely size distribution of particles. For the burning of coal, 79 percent of the particle emissions from electric generating plants and 62 percent from industrial sources were assumed to have been collected by control devices (NAPCA, 1970).
d Emissions are those under 3,000 feet only. Twenty percent of the fuel used in each flight is assumed to be consumed under 3,000 feet.

More detailed data can be found in Northern Research and Engineering Corp., 1969.
e N = Negligible.
f For some greater detail, see the report of Work Group 5 in this book.
g Includes such sources as forest fires, structural fires, coal refuse, agriculture, organic solvent evaporation, and gasoline marketing.

Work Group report and indicates that in 1967 some 13.4 billion metric tons of CO_2 were released from fossil fuel combustion and that emissions in 1980 (using Darmstadter's projection) would be 26 billion metric tons for the world as a whole.

As far as pollutants such as sulfur oxides, nitrogen oxides, hydrocarbons, carbon nonoxide, and particles are concerned, virtually nothing can be said with any assurance about their worldwide emission rates from energy production. The Division of Air Quality and Emissions Data of the National Air Pollution Control Administration (NAPCA) has calculated the U.S. emissions of these substances for the year 1968, and these figures are reproduced in Table 7.5. It is not possible to calculate worldwide emissions from the simple fuel-by-fuel breakdowns of world energy consumption that are now available, because the precise type and quantity of emissions associated with any energy-producing process depend upon innumerable variables in the fuel, the processing equipment, the scale and speed of operation, the effects of various types of control measures already being taken, the materials already present in the atmosphere, and a wide variety of other factors. The knowledge that would be needed to make even a theoretical calculation of world emissions of these substances does not exist, nor do the detailed empirical emissions studies from which one might determine worldwide emissions. (Or, if the data do exist, they do not appear to be available at a central information source.)

The problem of emissions of radionuclides is a special and very complex one. Probably the major potential for nuclear contamination of the environment will occur at the site of the fuel reprocessing plants. Here as the protective claddings and shields are removed to enable fuel recovery, fission and activation products are exposed and the potential for escape into the environment is increased. In fact, one estimate is that 99.9 percent of all radionuclides entering the environment are released from fuel reprocessing plants (Rivera-Cordero, 1970). A number of potentially hazardous radionuclides—such as iodine-151, zenon-153, strontium-90, and cesium-137—will have to be carefully monitored and controlled in order to limit the hazard to the public in the region of the plant. Possible releases of tritium (hydrogen-3) and of krypton-85 in particular are also of potential concern. In fact, considering the possibilities of local concentration effects due to weather, downwind conditions, and so forth, cleanup or containment procedures for krypton-85 will need to be developed (or perfected, since some work is already being done) substantially prior to the year 2000.

In Table 7.6 the radionuclides of potential contamination importance as liquid effluents are listed, along with their half-lives and the concentration released per 1,000 megawatts electrical (MWe) capacity from a PWR (pressurized water reactor) nuclear power plant. The presently accepted maximum permissible concentrations (MPC) in unrestricted water bodies are shown for comparison. Except for tritium (hydrogen-3) all of these isotopes are below MPC by at least six orders of magnitude

Table 7.6
Liquid Releases from a 1,000 MWe PWR Reactor

Isotope	Half-Life	Discharge Concentration Microcuries/cc	MPC Microcuries/cc
H^3	12.3 years	3.8×10^{-6}	5×10^{-3}
Mn^{54}	300 days	0.8×10^{-15}	8×10^{-4}
Co^{58}	71 days	2.4×10^{-14}	4×10^{-4}
Co^{60}	5.2 years	2.9×10^{-15}	1×10^{-4}
Sr^{89}	50.5 days	9.8×10^{-15}	7×10^{-5}
Sr^{90}	27.7 years	3.0×10^{-16}	4×10^{-7}
Y^{90}	64.8 hours	3.4×10^{-16}	3.0
Y^{91}	57.5 days	1.7×10^{-14}	0.2
Mo^{99}	67 hours	0.7×10^{-11}	8×10^{-4}
I^{151}	81 days	0.6×10^{-11}	2×10^{-4}
Cs^{134}	2.3 years	2.1×10^{-12}	9×10^{-6}
Te^{152}	78 hours	0.1×10^{-12}	5×10^{-4}
I^{133}	20.5 hours	0.7×10^{-11}	9×10^{-4}
Cs^{136}	13 days	0.9×10^{-13}	9×10^{-5}
Cs^{137}	27 years	3.4×10^{-12}	2×10^{-5}
Ba^{140}	12.8 days	2.3×10^{-15}	5×10^{-4}
La^{140}	40.5 hours	2.1×10^{-15}	2.0
Ce^{144}	290 days	0.8×10^{-14}	3×10^{-3}

Source: Wright, 1970.

(for cesium-134) and in most cases by more than eight orders of magnitude. Since even the most optimistic projections of nuclear power do not appear to exceed 10^7 MWe by the year 2000, even with no improvements in design and purification procedures of the liquid effluent, except for tritium, no radionuclide will be within 10^2 of the MPC. It should be noted, however, that there may be local buildups of specific radionuclides due to multiple power plants in a region and concentrations of the radionuclides in food chains.

Table 7.7 shows the estimated concentration factors for some radionuclides in aquatic organisms that were constructed from field studies on the Columbia River (Hanford reactor site) and the White Oak Lake (Oak Ridge reactor site). Other examples

Table 7.7
Estimated Concentration Factors in Aquatic Organisms

Radionuclide	Site	Phyto-plankton	Fila-mentous Algae	Insect Larvae	Fish
Na^{24}	Columbia River	500	500	100	100
Cu^{64}	Columbia River	2,000	500	500	50
Rare earths	Columbia River	1,000	500	200	100
Fe^{59}	Columbia River	200,000	100,000	100,000	10,000
P^{32}	Columbia River	200,000	100,000	100,000	100,000
P^{32}	White Oak Lake	150,000	850,000	100,000	30–70,000
Sr^{90}-Y^{90}	White Oak Lake	75,000	500,000	100,000	20–30,000

Source: Eisenbud, 1963.

from these areas include concentration factors of 1,000 for cesium in the flesh of bass, of 8,700 for zinc in the bones of the blue-gills, of 350,000 for radioactivity content in caddis fly larvae, 40,000 for duck-egg yolks, and 75,000 for adult swallows. Phytoplankton also tend to concentrate activation products such as zinc-65, cobalt-57, iron-55, manganese-54 to an even greater extent than fission products. For example, 88 percent of the radioactivity found in tuna from the open sea near the Marshall Islands was due to zinc-65, with iron-55 accounting for most of the remainder (Eisenbud, 1963).

In addition, though it is not an environmental pollution problem associated with nuclear reactors today, if and when the breeder reactor comes into common usage, careful consideration will have to be given to containment and monitoring for plutonium. Finally, there is the problem of perpetually managing the concentrated and highly radioactive wastes that are accumulated as a result of the use of nuclear power. Relevant data on the magnitude of this problem are contained in Tables 7.8 and 7.9. The projections of installed nuclear capacity used in those tables are similar to other projections summarized in the Battelle study and can be considered reasonable working hypotheses.

We feel that the problems associated with radioactive wastes deserve more careful study than we have given them. In particular, problems related to long-term storage of high-level wastes should be examined.

Table 7.8
Radioactive Wastes as a Function of Expanding U.S. Nuclear Power

	Calendar Year		
	1970	1980	2000
Installed nuclear capacity, MW(e)	11,000	95,000	734,000
Volume high-level liquid waste[a,b]			
Annual production, gal/yr	23,000	510,000	3,400,000
Accumulated volume, gal[c]	45,000	2,400,000	39,000,000
Accumulated fission products, megacuries[b]			
Sr^{90}	15	750	10,800
Kr^{85}	1.2	90	1,160
H^3	0.04	3	36
Total for all fission products	1,200	44,000	860,000
Accumulated fission products, tons	16	388	5,350

Source: Snow, 1967.
[a] Based on 100 gallons of high-level acid waste per 10,000 thermal megawatt days (MWd) irradiation.
[b] Assumes 3-yr lag between dates of power generation and waste production.
[c] Assumes wastes all accumulated as liquids.

7.4
Geographic Distribution of Emissions

We are aware that it would be of great use to meteorologists, ecologists, and oceanographers to know with some precision how these pollutants will be distributed around the globe. Unfortunately, existing energy projections and emissions data do not provide an adequate basis for such estimates. Only the distribution of past consumption of energy resources can be found in the U.N. data, and the worldwide projection cited earlier similarly offers a region-by-region breakdown solely of overall energy consumption for each fuel. This is enough to give some indication of the likely distribution of both heat and carbon dioxide emissions from fuel combustion and of the production of radionuclides, but little more.

Table 7.9
Waste Management Data for Conversion-to-Solids Concept

	Calendar Year			
	1970	1980	2000	(2030)[a]
Solid waste production, ft³/yr[b]	230	3,300	34,000	(——)
30-yr interim solid storage				
Volume in storage, ft³	450	16,000	310,000	(——)
Length 48-ft-wide canals, ft	6	177	3,420	(——)
1000-mile shipment to salt mine[c]				
Number of shipments per year	0	0	6	(670)
Number casks in transit	0	0	1	(4)
Disposal in salt mines				
Area required, acres/yr	0	0	1.2	(130)
Accumulated area used, acres	0	0	2.1	(1,185)

Source: Snow, 1967.
[a] Commitments made in year 2000.
[b] One cubic foot solid waste per 10,000 MWd; 3-yr lag between power generation and waste production.
[c] Thirty-six, 6-in.-diameter cylinders per shipment cask.

7.5

Potential of Pollution Control

It was found to be impossible to generalize about the potential effects of pollution control technology. It is clear, for example, that some pollutants (for example, the lead released by automobile engines) can be eliminated relatively simply and at a known cost. More expensive steps (for example, the shift from internal- to external-combustion engines in autos) would eliminate many more pollutants, but the economic factors and effects involved have yet to be even partially agreed upon. There is also the possibility of a greatly increased use of nuclear power that could help cut down certain emissions or even eliminate them over the longer run, but would produce still other emissions of potential concern as well as a complicated waste storage problem. And, of course, not even nuclear energy would solve a problem such as the generation of heat from energy production. Involved here are important questions of the proper mixes of energy production modes and control levels to be adopted so as to maximize society's overall gain from the necessary trade-offs of costs and benefits—questions that are quite beyond the capacity of any small group to deal with in a short time.

Greater control of various emissions is certainly possible, at a price. The present levels of control can be justified if society has made at least a quasi-conscious decision that the pollution effects are an acceptable cost for the specific benefits received. Decisions of this sort, however, are too often made implicitly and without any knowledge of the costs and benefits involved. The first priority for all concerned with the quality of life is to force the market and the political process to address

Table 7.A.1
World Production of Fossil Fuels, 1950–1967
(Millions of metric tons)

Year	Coal	Lignite[a]	Refined Oil Fuels[b]	Natural Gas[c]
1950	1,340	530	445	155
1951	1,375	550	530	180
1952	1,375	550	585	200
1953	1,380	555	605	210
1954	1,375	550	635	220
1955	1,500	630	705	240
1956	1,595	665	770	260
1957	1,625	765	795	285
1958	1,665	825	830	305
1959	1,730	845	900	345
1960	1,810	875	970	375
1961	1,625	905	1,030	405
1962	1,675	905	1,115	440
1963	1,740	965	1,210	480
1964	1,800	1,005	1,315	525
1965	1,815	1,030	1,410	565
1966	1,845	1,050	1,525	610
1967	1,750	1,040	1,630	655

Source: United Nations, *World Energy Supplies*.
[a] After 1962, lignite production figures were given in metric tons coal equivalent. For 1960, 1961, and 1962 (the only years for which there is overlapping data), the apparent conversion factor is 1 metric ton lignite = 0.44 metric ton coal equivalent. This same factor was used in the later data.
[b] Includes natural gasoline.
[c] Assumed density of 8×10^{-4} g cm^{-3} (1,000 m^3 = 0.8 metric ton) (President's Science Advisory Committee [PSAC], 1965)

Table 7.A.2
CO₂ Produced by Fossil Fuel Combustion, 1950–1967
(Billions of metric tons)

Year	Coal[a]	Lignite[b]	Refined Oil Fuels[c]	Natural Gas[d]	Total
1950	3.7	0.9	1.4	0.4	6.4
1951	3.8	0.9	1.7	0.5	6.9
1952	3.8	0.9	1.8	0.5	7.0
1953	3.8	0.9	1.9	0.5	7.1
1954	3.8	0.9	2.0	0.6	7.3
1955	4.1	1.0	2.2	0.6	7.9
1956	4.4	1.1	2.4	0.7	8.6
1957	4.5	1.3	2.5	0.7	9.0
1958	4.6	1.4	2.6	0.8	9.4
1959	4.8	1.4	2.8	0.9	9.9
1960	5.0	1.4	3.1	1.0	10.5
1961	4.5	1.5	3.3	1.0	10.3
1962	4.6	1.5	3.5	1.1	10.7
1963	4.8	1.6	3.8	1.2	11.4
1964	5.0	1.7	4.2	1.3	12.2
1965	5.0	1.7	4.5	1.5	12.7
1966	5.1	1.7	4.8	1.6	13.2
1967	4.8	1.7	5.2	1.7	13.4
1980 (est.)[e]	11.1[f]		10.8	4.0	26.0

Source: Computed from Table 7.A.1

[a] Assumed carbon content, coal = 75 percent (PSAC, 1965). Note: the weight of CO₂ emissions is equal to the weight of the fuel × its percent carbon × the ratio of the molecular weight of CO₂ to the molecular weight of carbon (i.e., $\frac{44}{12}$ = 3.67).

[b] Assumed carbon content, lignite = 45 percent (PSAC, 1965).

[c] Assumed carbon content, refined oil fuels = 86 percent (the figure used by PSAC, 1965, for liquid hydrocarbons).

[d] Assumed carbon content, natural gas = 70 percent (corresponding to a mixture by volume of 80 percent CH₄, 15 percent C₂H₆, and 5 percent N₂) (PSAC, 1965).

[e] The 1980 estimate was constructed by multiplying the 1965 emissions for coal (including lignite), refined oil fuels, and natural gas × growth factors (i.e., ratios of 1980 consumption to 1965 consumption for solid and liquid fuels and natural gas) derived from Darmstadter's figures in metric tons coal equivalent for 1965 and 1980.

[f] Coal and lignite combined.

the questions squarely and explicitly, and to see that the data necessary for accurate responses are quickly and reliably collected.

7.6

Appendix: Calcuation of CO_2 Emissions

This calculation of global emissions of carbon dioxide is modeled on (and updates) the one contained in Appendix Y4 of *Restoring the Quality of Our Environment* (President's Science Advisory Committee [PSAC], 1965) The only change in this present calculation is the use of production figures for all refined oil fuels rather than figures for crude petroleum production which the PSAC calculations employed.

This change is consistent with the practice in the U.N. source of considering only refined oil fuels in the tabulations of the consumption of energy from liquid fuels (United Nations, *World Energy Supplies*). The U.N. figures for the production of refined oil fuels exclude nonenergy products (that is, lubricants, bituments, paraffin wax, road oil, petrochemical feedstocks, and petroleum coke). Therefore, production figures for all refined oil fuels provide a better basis for calculating CO_2 emissions from the combustion of liquid hydrocarbons.

Table 7.A.1 presents world production figures for fossil fuels and Table 7.A.2 the emissions of CO_2 that would result from the combustion of these fuels.

References

Battelle, 1969. *A Review and Comparison of Selected United States Energy Forecasts*, prepared for Office of Science and Technology by Pacific Northwest Laboratories of Battelle Memorial Institute (Washington, D.C.: U.S. Government Printing Office).

Darmstadter, J. *Energy and the World Economy* (Baltimore: The Johns Hopkins Press for Resources for the Future, Inc.), forthcoming.

Eisenbud, M., 1963. *Environmental Radioactivity* (New York: McGraw-Hill). *Minerals Yearbook*, 1968 (Washington, D.C.: U.S. Bureau of Mines, U.S. Government Printing Office), 1969.

National Air Pollution Control Administration (NAPCA), 1970. *Nationwide Inventory of Air Pollutant Emissions* (Raleigh, North Carolina: NAPCA).

Northern Research and Engineering Corp., 1969. *Nature and Control of Aircraft Engine Exhaust Emissions*, NAPCA Contract No. PH22–68–27 (Cambridge, Massachusetts).

President's Science Advisory Committee (PSAC), 1965. *Restoring the Quality of Our Environment* (Washington, D.C.: U.S. Government Printing Office).

Rivera-Cordero, A., 1970. The nuclear industry and air pollution, *Environmental Science and Technology, 4.*

Seaborg, G., 1969. Testimony in *Environmental Effects of Producing Electric Power*, Hearings before the Joint Committee on Atomic Energy, 91st Congress, 1st Session, Part I (Washington, D.C.: U.S. Government Printing Office).

Snow, J. A., 1967. Radioactive waste from reactors: the problem that won't go away, *Scientist and Citizen, 9.*

United Nations. *World Energy Supplies*, Statistical Papers, Series J (New York: Statistical Office, Department of Economic and Social Affairs, United Nations).

Wright, J. H., 1970. Power and the environment (paper presented at the American Power Conference, Chicago, Illinois, April 21–23, 1970).

Units and Conversion Factors

Weight
1 kilogram (kg) = 1,000 grams (g) = 1,000,000 milligrams (mg) = 1,000,000,000
   micrograms (μg)
1 kg = 2.205 pounds (lb); 1 g = 0.035 ounce (oz)
1 lb = 453.6 g = 0.4536 kg
1 metric ton = 2,205 lb = 1,000 kg
1 short ton = 2,000 lb = 907.2 kg
1 megaton = 1,000,000 tons

Length
0.001 kilometer (km) = 1 meter (m) = 100 centimeters (cm) = 1,000 millimeters
   (mm) = 1,000,000 microns (μ)
1 km = 0.6214 statute mile; 1 m = 39.37 inches (in.) = 3.281 feet (ft); 1 cm
   = 0.3937 in.
1 mile = 1.609 km; 1 ft = 0.3048 m; 1 in. = 2.54 cm

Area
1 hectare = 10,000 square meters (m^2)
1 hectare = 2.47 acres = 0.003861 square mile
1 square mile = 640 acres = 259 hectares

Volume and Cubic Measure
1 cubic meter (m^3) = 1,000,000 cubic centimeters (cm^3)
1 m^3 = 35.31 cubic feet (ft^3); 1 cm^3 = 0.061 cubic inch ($in.^3$)
1 liter (l) = 1,000 cm^3
1 l = 61.02 $in.^3$ = 0.2642 gallon (gal)
1 ft^3 = 0.02832 m^3 = 28.32 l; 1 gal = 231 $in.^3$ = 3.785 l

Energy and Work
1 British thermal unit (Btu) = 252 calories (cal) = 0.0002931 kilowatt-hour
   (kWh)
1 kWh = 860,421 cal = 3,412 Btu

Power
1 megawatt (MW) = 1,000 kilowatts (kW) = 1,000,000 watts (W) = 3,413,000
   Btu/hour (Btu/hr) = 1,341 horsepower (hp)

Pressure
1 atmosphere (atm) = 76 cm mercury = 14.70 $lb/in.^2$ = 1,013 millibars (mb)

Speed

Mach number = $\dfrac{\text{speed of aircraft}}{\text{speed of sound in surrounding atmosphere}}$

At Mach number 1.0 (Mach 1.0), the speed of an aircraft is about 760 miles/hr
   (mph) = 1,223 km/hr (at sea level and room temperature)
At Mach 2.0, the speed is 1,520 mph = 2,446 km/hr

Temperature scales

	Absolute Zero	Ice Point (water)	Steam Point (water)
Degrees Fahrenheit (°F)	−459.7	32	212
Degrees Celsius or Centigrade (°C)	−273.15	0	100
Degrees Kelvin (°K)	0	273.15	373.15
Degrees Réaumur (°R)		0	80

Index